nature's robots

nature's robots

A history of proteins

CHARLES TANFORD &
JACQUELINE REYNOLDS

Emeritus Faculty, Duke University

OXFORD

UNIVERSITY PRESS

Great Clarendon Street, Oxford OX2 6DP

Oxford University Press is a department of the University of Oxford.
It furthers the University's objective of excellence in research, scholarship,
and education by publishing worldwide in

Oxford New York

Athens Auckland Bangkok Bogotá Bombay Buenos Aires Calcutta
Cape Town Dar es Salaam Delhi Florence Hong Kong Istanbul
Karachi Kuala Lumpur Madrid Melbourne Mexico City Mumbai
Nairobi Paris São Paulo Singapore Taipei Tokyo Toronto Warsaw

and associated companies in
Berlin Ibadan

Oxford is a registered trade mark of Oxford University Press
in the UK and in certain other countries

Published in the United States
by Oxford University Press Inc., New York

© Charles Tanford and Jacqueline Reynolds, 2001

A catalogue record for this book is available from the British Library

Library of Congress Cataloging in Publication Data

ISBN 019 850466 7

Typeset by Footnote Graphics, Warminster, Wiltshire
Printed in Great Britain
on acid-free paper by
T.J. International,
Padstow, Cornwall

Acknowledgements

We gratefully acknowledge the support and encouragement we have had from John Edsall and Walter Gratzer and from Michael Rodgers of Oxford University Press. We also thank a number of our contemporaries who have contributed authoritative factual knowledge or some of their own recollections: Robert Baldwin, Jim Huston, Walter Kauzmann, Darrel McCaslin, Herbert Morawetz, Hans Neurath, Jens Nørby, Max Perutz, and John Schellman. We are grateful to George Gaines, Fred Sanger, and Giorgio Semenza for photographs from their personal collections.

Special appreciation is due to the staff of the Science Museum Library, London, for their cheerful retrieval of many volumes of old scientific publications from the depths of the archival storage area.

Contents

Part IV How are proteins made?

Introduction

In September 2000, about the time that the manuscript for this book was being completed, the National Institute of General Medical Sciences, part of the US National Institutes of Health, launched the largest explicit molecular project of all time, aiming to solve the three-dimensional structures of 10 000 proteins. The goal is to do this not for just any 10 000 proteins, but to select ones that would represent identifiable *protein families*, each family representing a group with similar physiological function and/or gene similarities—proteins expected to be sufficiently closely related so that knowledge of one structure can be expected to allow prediction of structures for other members of a family on the basis of much less information than rigorous atom-by-atom measurements. The project is known as the 'structural genomics initiative'. The work is to be split among seven regionally based research groups. The cost to the institute over the first five years will be $150 000 000 (£100 000 000).

The launching of this project provides a fitting climax to our history of protein science, which takes us from the origins of protein research in the nineteenth century, when the chemical constitution of 'protein' was first studied and heatedly debated and when there was as yet no glimmer of the functional potential of substances in the 'protein' category, to the determination of the first structures of individual proteins at atomic resolution—when positions of individual atoms were first specified exactly and bonding between neighbouring atoms defined precisely. The numerical explosion of such detailed information from a handful to 10 000 proteins is something we could not have imagined, nor do we intend to dwell on it. Our objective is limited to history, the heroes of the past, who worked mostly alone or in small groups, usually with little support from formal research grants. How did we get from scratch to where we are? That is our question, rather than prediction of the future and the huge range of possibilities now opening up, the brave new world that looms on the horizon. No portents of things to come—we read the

Sunday papers for that. It will be 20 or 30 years before the time is ripe to write the history of the most recent period.

The history of protein science as a whole, of course, includes subject matter beyond the specifics of three-dimensional atomic organization. The most widely appreciated is physiological function and how it might be related to molecular structure, for proteins are amazingly versatile molecules. They make the chemical reactions happen that form the basis for life, they transmit signals in the body, they identify and kill foreign invaders, they form the engines that make us move, they record visual images. All of this is common knowledge, but it wasn't so a hundred years ago. Even the knowledge that proteins are the responsible agents for many biological functions is not really ancient. As for the relation between structure and function, it is not usually self-evident from mere inspection of structure, even today, when so many structure/function relations have in fact been conclusively worked out.

In treating this topic we again emphasize the pioneering efforts of the early days. How did we get from recognition in 1902 that sodium and potassium ions trigger the electrical impulse of a nerve signal to the modern discovery of the responsible proteins, the sodium/potassium pump and the cation channels? How did we get from recognition (already in 1802!) that perceived colours can be decomposed into three primary components to the modern discovery of three matching protein receptor molecules? How, in muscle contraction, did we progress from the intuitive idea that some molecular component of muscle must contract (helix to coil, something like that) to the modern sliding filament model, which is based on molecules that are constant in length and achieve contraction by sliding past each other? In immunology, how did we progress from recognition of the existence of specific antisera to an understanding of the basis for achieving the requisite variety in molecular structures?

The history of protein science as a whole must also deal with protein *synthesis* in the living organism. Here one is in a somewhat different position as compared to protein structure or its relation to physiological function, for the elucidation of the manner of protein synthesis has taken place mostly in the last four decades and has involved a merging of protein science with molecular biology—with its well-known dramatic applications to all aspects of life and evolution and ramifications going far beyond the scope of our book. We have treated this topic in three chapters at the end of our book, but we must emphasize that they are brief, intended only to provide historical perspective to the question: How are proteins made?

Comment on robots

We deliberated a long time about the title of this book. Is the word 'robot' really appropriate? In one sense the answer is certainly yes: robots are automatons—you don't need to tell them what to do, they already know. Proteins satisfy that criterion. For every imaginable task in a living organism, for every little step in every imaginable task, there is a protein designed to carry it out. And it is programmed to know when to turn on or off. The ultimate objective of a task may be chemical or mechanical or to measure colour or fight off a foreign invader—there is no limit to what can be accomplished. The common feature is that proteins are in control and know what to do without being told by the conscious mind.

Of course, for most of us, what springs to mind when we hear the word 'robot' is an articulated mechanical device, designed by human ingenuity and often even with human features. But it seems to us not unreasonable to extend the image to molecules, designed by genes, as proteins are. Richard Dawkins has (famously) coined the term 'selfish gene'. Perhaps it is appropriate to think of proteins as the instruments whereby all genetic programmes are automatically carried out and there is little or nothing that the organism as a whole—its conscious mind if it has one—can do to alter their individually programmed functions.

The language of chemistry, with special reference to the general reader

Proteins are made up of molecules and, like all molecules, they are uniquely defined by their constituent atoms and by how these atoms are linked to each other and (in many cases) by how they interact with other molecules around them, such as the water molecules in the typical biological environment. In fact, proteins are somewhat special in that they are in the category of *macromolecules*, giant assemblies with many thousands of atoms apiece, dwarfing most other kinds of organic molecules with which the general reader is likely to be familiar—even those with what often seem to be magic powers, like steroids, penicillin, vitamins, etc., are vastly smaller on a molecular scale. The recognition of the macromolecular nature of proteins turns out to be a pivotal point in the history of protein science; for a long time there were people who did not believe that such large molecules, with precisely defined atomic composition and structure, could exist.

To discuss this question of molecular size (even to define it), we need of course to make use of the language of chemistry. When we go beyond bare descriptive definition of molecules, to all aspects of how proteins behave, it is inevitable that observations, questions, and answers will likewise need to be couched in the language of chemistry—every imaginable function depends on a protein molecule's atoms and how they are arranged. Such terms as 'complementarity', 'binding sites', 'recognition' between hormones and cellular receptors, etc., are clearly comprehensible in themselves, but, when used explicitly to focus on a particular protein, a particular robotic function, then a protein's atoms inevitably take centre stage and we find ourselves grappling with chemical language again. What this means overall is that a reader with some background in chemistry or in other disciplines that tend to use the language of chemistry (for example, pharmacology or physiology) may find himself or herself more at ease with parts of our text than someone who lacks previous exposure to the subject. One might go so far as to say that biochemists and their close relatives can reasonably be expected (on the basis of this familiarity) to constitute our primary audience—our book is likely to find its resting place in the reader's library on the same shelf as histories of biochemistry.

But, pointing this out (in a sense belabouring the obvious) provides the opportunity to emphasize that the chemistry of proteins is in fact quite simple, in spite of the giant molecular size. Only five different atoms—carbon, hydrogen, nitrogen, oxygen, and sulphur—are used in constructing the majority of protein molecules. Some proteins do contain other essential atoms (notably metal atoms), but even then these five kinds of atom alone create the ubiquitous central molecular fabric, the core of every single protein that needs to be mentioned in the historical context as contributing to an understanding of proteins. To simplify the chemistry even more, these five atoms, intrinsically capable of existing in a virtually infinite variety of mutual linkages, occur in the central protein *exclusively as amino acids*, a mere handful of essentially similar combinations, and the amino acids are always joined to one another in the simplest possible manner: linearly in long chains.

Again, some proteins contain other atomic groupings, not related to amino acids, but they are usually treated as structurally supplemental to the ubiquitous chains built from amino acids: the jargon for groupings like the porphyrin ring central to the haem group of haemoglobin is the term 'prosthetic group'—which does not imply a lack of importance, but does suggest that consideration of the chemical details can be temporarily put aside in the quest for a fundamental understanding of the *common features of all proteins*. And this is historically how such supplemental groups have been approached, even in the case of the haem group of haemoglobin, which includes the vital iron atom that enables the protein to carry oxygen.

This leaves the purely chemical story of proteins focused primarily on the amino acids, few in kind, as we have already said, and always combined into long linear chains. How are they linked? How many in a given chain? How are the chains folded in three-dimensional space? What sort of influence and strains do individual amino acid units impose upon the environment? Because the same amino acids are involved over and over again (and this will be true even in the thousands of proteins still envisaged for future investigation) it will prove to be relatively simple. It is all far less daunting than the chemistry of many other systems that one encounters in daily life. Look, for example, at the information leaflets that are usually packaged with medicines that you buy at the pharmacy: they reflect specialization *ad extremum*, with no discernible connection between one chemical substance and another.

Overall purpose

With the foregoing preamble in mind, we can proceed to define our objectives in this book. We start from the premise that proteins are at the heart of all living processes, uniquely versatile in their capability to do whatever is needed, and with the knowledge that proteins are increasingly being harnessed to serve practical needs of society, to cure disease, to safeguard crops, etc. As a result, proteins are now moving to centre stage in the theatres of medicine and biology; scientists with specialized training in protein science are increasingly in demand in both academic and industrial environments. These new recruits, if past experience is a valid guide, will be focused on tasks ahead and their historical perspective will usually be dim, and many of their older, more experienced colleagues may still be in the same situation. These active protein scientists of the present day are probably the primary target for our book—we want to tell them that their work rests on the shoulders of predecessors, some well-known, some less so, not all of them by any means among the legendary 'giants' who tend to dominate popular histories. We seek to identify the pioneers and to define their achievements.

But we hope that our readership will go beyond those already committed to protein science by professional ties. We sense that the greater publicity (the being 'centre stage') enjoyed recently by biological science in general and proteins in particular has made the general reader much more aware of the subject than in the past. With this in mind, we aim to present successive discoveries in protein science in a manner that will be comprehensible to the general reader, but to do so without dumbing down the chemical content, without short-changing the scientist in the laboratory. We shall put in occasional footnotes where we think that they might help the novice appreciate some chemical nuance that might otherwise be missed, but this won't happen very often—there is no harm done if the reader does not know the actual formula of every single substance that is mentioned; even professional chemists would have to look most of them up. In any case, we don't spell out most formulas: we are not writing a textbook and even the amino acids, the absolutely essential constituents of proteins, are listed only by name. (And most of the names have actually found their way into standard household dictionaries, such as Chambers or the Shorter Oxford.)

Our overall advice to the general reader: don't be intimidated by chemistry! Only a minimum of technical language is needed to get a feeling for the growing excitement that was generated over the years as the mysteries of protein structure and function—the core of all the mysteries of life—were revealed step by step.

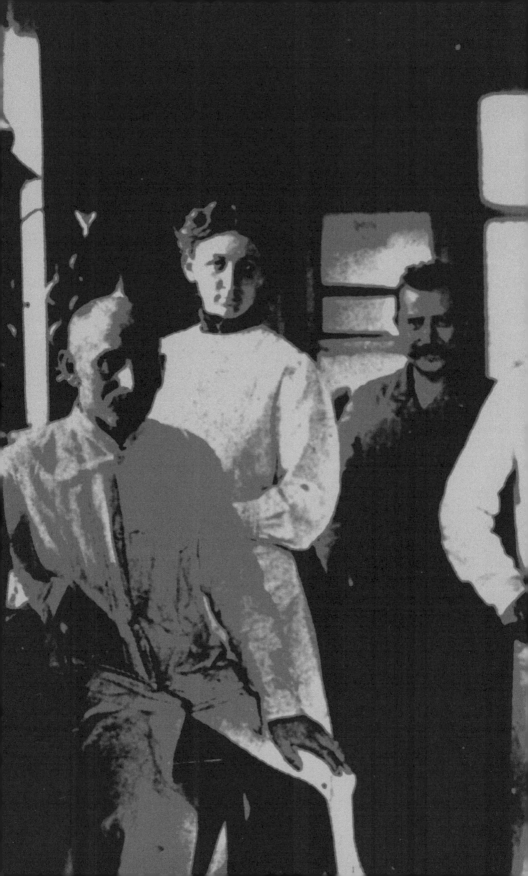

PART I

Chemistry

CHAPTER 1

The naming

Le nom protéine que je vous propose pour l'oxyde organique de la fibrine et de l'albumine, je voulais le dériver de πρωτειοζ, parce qu'il paraît être la substance primitive ou principale de la nutrition animale
Letter from J. Berzelius to G. Mulder, 10 July, 1838[1,2]

Substances we now call proteins aroused the interest of early-day chemists because of their close association with life processes. The French chemist Antoine Fourcroy (1755–1809)[3] recognized three distinct varieties of protein from animal sources in 1789: albumin, fibrin, and gelatin. At least two more were known from vegetable sources. They were often collectively called 'albumins', in recognition of the prototype derived since time immemorial from egg white; the German equivalent was 'Eiweisskörper'. Later on (after 1800), when the distinction between elements and compounds was sorted out and analytical methods for elemental composition became reliable, these substances, along with other natural products, became targets for quantitative study—publications galore from all over Europe, seeming to consist of little more than endless tables of percentages: carbon, hydrogen, nitrogen, oxygen, sometimes sulphur or phosphorus as well. Compositional formulas were often given, based on the data and a scale of atomic weights, and formula weights were sometimes calculated. But, after the work was done, there was little certainty about the meaning of the results in the modern sense. What could a compositional formula mean when bonding between atoms was still not even vaguely understood?

There are many modern historical accounts of this formative period in the science of organic chemistry, to which the reader is directed for more detail.[4–7] It suffices to say here that the big step forward—from a poorly defined conglomeration of atoms to an organized molecular model with explicit interatomic bonds—did not come until after the tetravalence of carbon was proposed by August Kekulé (and generally accepted) around 1857.[8] This must be kept in mind as this initial chapter of our book is read: the recognition of the special place of proteins came during what were still the dark ages for organic chemistry as a whole.

1838: Jacob Berzelius and Gerrit Mulder

The principal characters in the story of the naming are the Swedish chemist Jacob Berzelius and the much younger Gerrit Mulder, Dutch physician and chemist, one of the 'analysts'. The manner of interaction between them provides a vivid portrait of the time, as it affected scientific communication, perhaps all the sharper for being seen explicitly as part of the protein story,[9] rather than as part of a more ponderous tome on communication as a whole.

Jöns Jacob Berzelius (1779–1848) was by 1838 almost 60 years old, the elder statesman of the world of chemistry.[10] He had had a difficult path to his prominent position, often in financial trouble and never with any decent laboratory space or equipment until around 1810, when he became a professor at what is now the Karolinska Institute in Stockholm. His achievements in chemistry are therefore all the more noteworthy. He discovered several new chemical elements: Li, Se, V, Ce, Th; he assigned definitive atomic weights to most of the atoms then known, proposed highly influential theories of interatomic attraction, and wrote several textbooks, including one on animal chemistry, which was not as widely read as Justus Liebig's later book with the same title. In 1812 he became a member of the Swedish Academy of Science and shortly thereafter was named its president.

What is most remarkable about Berzelius, as we peruse his works today, is that he was a prodigious correspondent, receiving and writing letters from his desk in Stockholm, many volumes of which have been published: there is correspondence with notables from abroad, such as Wöhler, Berthollet, Humphry Davy, Mulder, and Mitscherlich; and many less familiar names appear, including a number of fellow-Swedes.[1,11] It was his way of keeping himself informed, interacting with other scientists, at a time when (at least in Sweden) there were virtually no opportunities for groups of chemists to gather in one place; even personal contacts usually had to be one-on-one, and tended to involve difficult travel. In a letter written to Mulder in 1834, Berzelius remembers a journey in 1828 when he was actually in Rotterdam for a few hours and would have enjoyed a talk with Mulder had he known him then—the journey was a short one, from Antwerp to The Hague, but it involved a sailing ship as far as Rotterdam and then an overland stagecoach from there, for which he had had to wait.[12] The letters extend over most of Berzelius's career, from 1804 until his death; the correspondence with Mulder (76 letters) began in 1834. The letters were, of course, hand-written and French was the usual language when the correspondent was not Swedish. Berzelius and Mulder both used French; Davy's letters were in English and Berzelius replied in French; Eilhard Mitscherlich, professor in Berlin, wrote in German and Berzelius replied in Swedish—Mitscherlich, a former Berzelius student, was presumably reasonably proficient in Swedish.

G. J. Mulder (1802–1880) had an easier start in life than Berzelius. His father was a surgeon and could afford a regular degree course in medicine for his son. Mulder practised medicine and at the same time dabbled in chemistry, but had to give up medicine when the demands made on him by the cholera epidemic of 1832–1833 proved too arduous. Stable university faculty appointments followed. Mulder was at first a lecturer at Rotterdam, in 'botany, chemistry, mathematics and pharmacy', all of them presumably simultaneously. In 1840 he became professor at Utrecht, where he remained until his retirement. He was an energetic man, ultimately more active as teacher, writer of textbooks, and editor of journals, than in his laboratory research. He lectured three hours a day to large classes. He had 10 or 20 students in his private laboratory, most of them there for training rather than research.

In his own laboratory work, Mulder was, as we have said, one of the 'analysts', producing compositional data for any natural product he could get hold of, including substances from plants that were reputed to have medicinal value—for example, extracts from the bark of trees and natural oils. There was much discussion in the Berzelius correspondence in1836 and 1837 of Mulder's analyses of tea and coffee: four kinds of tea, two from China and two from Java. But in 1837–1838 Mulder began to concentrate on the elemental analyses of the nitrogen-containing substances that were then designated as 'albumins': fibrin, egg albumin, blood serum albumin, and wheat albumin—essentially those defined at the time of Fourcroy fifty years earlier. He was guided by and heavily dependent on Berzelius, who in his letters gave advice on many topics; in particular, Berzelius suggested the methods of analysis for phosphorus and sulphur which these substances were prone to contain in addition to the more usual elements.

Fig. 1.1 Gerrit Mulder in ceremonial regalia. (Source: Science Museum/Science & Society Picture Library.)

And then an extraordinary result emerged unexpectedly from these analytical data and got Mulder tingling with excitement. Apart from the sulphur and phosphorus, these seemingly distinct albumins had strikingly similar compositions: in fact, they appeared virtually identical, as illustrated by the sample table in Fig. 1.2, reproduced from Mulder's published results.[13,14] We noted earlier that there was uncertainty about the meaning to be attached, in the modern sense, to quantitative compositional data. But to have virtually identical compositions for substances you expected to be different is another matter altogether and the significance of such a finding is unrelated to what it is that is holding the atoms together. The result led Mulder to a speculation that a single core substance, which he called 'Grundstoff', might be *identical* in all these

	Fibrin		Albumin		
			v. Eiern		v. Serum
Kohlenstoff .	54,56	—	54,48	—	54,84
Wasserstoff .	6,90	—	7,01	—	7,09
Stickstoff .	15,72	—	15,70	—	15,83
Sauerstoff .	22,13	—	22,00	—	21,23
Phosphor .	0,33	—	0,43	—	0,33
Schwefel .	0,36	—	0,38	—	0,68

Fig. 1.2 Table of Mulder's analytical results, reproduced from his paper in 1838.[14] The formula calculated by Mulder for fibrin or egg albumin, on the basis of the content of phosphorus and sulphur, was $C_{400}H_{620}N_{100}O_{120}P_1S_1$. For serum albumin it was the same, but with two atoms of sulphur.

substances with only added sulphur or phosphorus creating the differences. And this led to the further speculation that this 'Grundstoff' might in fact be synthesized in only one host (wheat or other plants) and then transferred intact as part of the process of nutritional assimilation into animals. To quote Mulder's own words:

> I hold it now as established that the main mass of animal matter is delivered directly from the plant kingdom, and that fibrin and albumin, except for 1 atom of sulphur, have the same composition. The vegetable 'Eiweiss' was synthesised from grain . . . .

These are strong words, sensational even. A single substance, appearing with but minor modification as every known protein in the world!

Mulder summarized his results for Berzelius in a letter (3 June 1838) and Berzelius became quite excited, too. A special distinctive name seemed appropriate and 'protein' was the name he suggested in his reply a month later. The name is derived from the Greek word 'πρωτειοζ' which means 'standing in front', 'in the lead'. 'We're number one' is how one might translate it in the modern vernacular. Berzelius justified the name in a letter to Mulder, dated 10 July 1838, on the basis that protein 'seems to be the primitive or principal substance of animal nutrition, which plants prepare for the herbivores, and the latter then furnish for carnivores'.

It is important to note that Mulder's theory was not really as unreasonable a flight of fancy as it might seem today, because it was consistent with the 'radical' theory, a new idea which was rapidly gaining popularity among chemists at the time, an idea that promised to create some order within the complexities of organic chemistry. It was based on the observation that certain groups of atoms seemed to be rather stable, occurring repeatedly and retaining identity through a series of reactions—in other words, sharing some of the attributes of single atoms. They were first called radicals by Gay-Lussac as early as 1815, but a lot of people endorsed the idea. Advocacy by Berzelius was particularly strong and he is generally regarded as a major influence in their acceptance into standard chemical doctrine. Mulder's 'Grundstoff' can be seen as an extension of the 'radical' concept, albeit of unprecedented size. Mulder's formula for his protein core was $C_{400}H_{620}N_{100}O_{120}$, much bigger than methyl, ethyl, etc.

Liebig's influence

Justus Liebig entered the scene almost immediately after the publication of Mulder's paper—predictably so, for Liebig, professor since 1824 at the University of Giessen, almost exactly the same age as Mulder but already as celebrated as Berzelius, had a huge ego and an aggressive aspiration to personal glory in all things chemical. Nothing of importance would have been allowed to escape his enthusiasm or venom—sometimes both, one right after the other. A recent biography has dubbed him the 'chemical gatekeeper', who opened communications between chemistry and related fields: nutrition, physiology, fermentation, and agricultural practice were among the subjects actively in his sights.[15]

Liebig wrote many books, most of them published simultaneously in German and in English translation. In 1839 there was a monograph on chemical analysis of organic compounds.[16] In 1840 came the first of his 'gateway' books, on the applications of chemistry to agriculture,[17] followed in 1841 by his equally famous and often reprinted *Animal chemistry*, subtitled *Organic chemistry applied to physiology and pathology*.[18] According to historian E. Glas,[19] *Animal chemistry* was written in a hurry, at a time when there was a huge upsurge in protein-related work throughout Europe, most of which was the direct result of Mulder's protein theory. The book had been intended to establish priority of concepts and is full of bold speculation without experimental proof, but nevertheless (quoting Glas), 'it had stimulated physiological chemistry greatly before it passed away'.[19] At the time of Mulder's historic paper, Liebig was also editor of *Annalen der Pharmacie*, without apparent restraints on what he could say in his own journal.[20] In 1839 he used this vehicle for the best-known of many examples of his venom, a savage satirical attack on the cell theory of living organisms, ridiculing the pioneering work of Theodore Schwann in Germany and Cagniard de la Tour in France, who had demonstrated that the 'ferments' which transform sugar into alcohol (for example, in wine making) were derived from a living yeast. Not so in Liebig's eyes, who firmly believed that all chemistry in vats and laboratory vessels had to be pure chemistry, without intervention of living organisms.[21]

Like Berzelius, Liebig had a vast correspondence with other scientists.[22] Mulder's historic paper was in fact part of a letter that Mulder wrote to Liebig, initiating another lengthy correspondence.[23] Liebig was aroused to enthusiasm and published the substance of the letter in his *Annalen*. He replied to Mulder in 1841, full of praise, saying that everything Mulder had done was right and had been confirmed in his laboratory. He was ecstatic about the idea that a plant protein could be transferred unaltered into the animal world and went on to convince himself that only four distinct proteins existed in all the world's plants.[24,25] As regards the animal proteins, he went far beyond Mulder in his *Animal chemistry* book, with wild claims, such as:

> Both albumen and fibrine, in the process of nutrition, are capable of being converted into muscular fibre, and muscular fibre is capable of being reconverted into blood.[26]

As it happened, Mulder's analytical results quickly came into question. In Liebig's own laboratory, Mulder's claim that sulphur could be separated from its parent proteins to yield the so-called 'Grundstoff' could not be substantiated—some sulphur always remained.[27-29] In Dumas' laboratory in France, small variations from protein to protein were found in the supposedly invariant atomic ratios—for example, a higher nitrogen content was found in fibrin than in egg albumin.[30] Liebig's admiration for Mulder turned overnight to unprincipled contempt and Mulder was made the scapegoat for Liebig's own questionable pronouncements. Mulder had *enticed* chemists to focus on elemental analysis, Liebig said in a blistering attack, when they should have been working on initial stages of decomposition and studying the products thereof.[31]

But the very existence of the Liebig–Mulder dispute enhanced the status of proteins. Liebig's earlier enthusiasm had enforced the Berzelius–Mulder notion of the unique importance of proteins—without Liebig it might have been less durable.*

But once planted in the mind, the importance, the sense of unlocking some secret of life, remained. It had seemed a miracle to be able to think of every known protein in the world as basically a single substance. But when it turned out that every protein was different, that dozens of them could be identified, similar but distinct, physiologists began to discern a link between unique proteins and unique physiological functions—in its own way this was even more of a miracle. The name 'protein' has quite appropriately continued in use, albeit for a long time in parallel with older terminology: 'Eiweisskörper' remained the standard nomenclature in Germany until the beginning of the twentieth century and one continues to find 'albuminous material' or slight variants of that expression in English publications. (But by 1906, when Otto Cohnheim's influential *Chemie der Eiweisskoerper* was translated into English, the title was changed to *Chemistry of the proteids*.[32])

Formula weight or molecular weight

An offshoot of Mulder's paper, a direct experimental fact, involving no extravagant theoretical ideas, is the demonstration that proteins must have exceptionally high molecular weights; they must be macro-molecules, to use the modern term.

The conclusion is based on the elemental analyses themselves, from which a minimal formula weight can always be derived by using the simple principle that no molecule may contain a non-integral number of

---

* Note that Liebig was controversial and often absurdly incorrect in many of his 'gatekeeper' efforts. Being widely regarded as an infallible oracle, his applications of organic chemistry to practical agriculture could have been quite damaging. For example, Liebig asserted that agricultural fertilizers did not need to contain nitrogen because plants could get all the nitrogen they needed for protein synthesis from the air. John Lawes, founder of the Rothamsted research station in England, was a vigorous challenger of Liebig's views.[33]

atoms. The derived formula weight will thus inevitably be large for *any organic compound* whenever at least one element is present in very small amount. The compound must of course be deemed pure and the analytical content of the critical element (the ratio to other more abundant elements) must be accurately reproducible from one experiment to another to provide convincing demonstration of being an integral part of the molecule. Likewise, the element must be accounted for among the products of chemical reactions that split the original molecule into fragments. These conditions are *uniquely* satisfied by proteins, which nearly always contain small amounts of sulphur, derived from the amino acids methionine and cysteine; moreover, proteins are easily broken down into fragments that satisfy the requisite compositional conservation. Other natural molecules that were being studied at the time do not have atomic constituents of low frequency. Cellulose, starch, and rubber, for example, are composed of C, H, and O alone, all at comparable atomic levels, and analysis (allowing for a reasonable margin for experimental error) will inevitably yield a plausible minimal molecular weight in the range of what we now know to be the monomer in a polymeric chain.

The formula given earlier for Mulder's 'Grundstoff'[14]—the portion combined with a single atom of sulphur or phosphorus—was $C_{400}H_{620}N_{100}O_{120}$, yielding a molecular weight of about 54 000, based on an atomic weight scale with O = 100. Several different atomic weight scales were in use in different laboratories at the time, some with C/H = 6, for example, but it happens that the *relative* atomic weights of Mulder's scale were correct, the same as they would be today, a fact not recognized in some accounts of the work.[33] Thus Mulder's formula weight can be directly translated to the modern scale with O = 16, simply by multiplying by 16/100. The result is a formula weight of 8600. This is a large number, without question in the 'macromolecule' category even by today's standards, and of course it is only a *minimum*, to be multiplied by an appropriate factor for proteins with several S atoms per molecule. Mulder's analytical techniques were evidently good, as evidenced by his measurement in the same paper for leucine: the minimal value, for one atom of nitrogen and O = 16, was 132, compared to the true molecular weight of 131. Mulder recognized that leucine is a decomposition product of proteins, which made its much smaller size entirely reasonable.[14]

The dispute with Liebig, mentioned earlier, in no way affects the argument for large intrinsic size. Liebig had no dispute with the presence of a small amount of sulphur in the proteins studied by Mulder (Fig 1.2). On the contrary, what Liebig objected to was Mulder's assertion that the sulphur could be easily removed. The sulphur content being the key analytical figure that inevitably leads to a high formula weight, then the harder the sulphur is to get rid of, the more confident we can be that it is an essential atomic constituent. This is borne out by progress in the ensuing years, when colloid chemists seriously believed that there could be no true molecules with molecular weights even as large as Mulder's 8600, as Chapter 4 of this book will discuss in detail. As criteria for protein purity and accuracy of elemental analysis got better and better, the sulphur remained, a bulwark for many decades of those who argued for the macromolecular concept.

In conclusion, to avoid confusion for readers who decide to sample some of the original literature, it should be noted that 'atomic weight' and 'molecular weight' were used more or less interchangeably in Mulder's time, the former implying the weight of the sum of the constituent atoms. Furthermore, there was probably not much distinction in the minds of many investigators between these weights and the term 'binding equivalent'. Since Mulder claimed to be able to remove sulphur and leave an intact protein core, he may have thought of the latter as the entity that *binds* one atom of sulphur.

CHAPTER 2

Crystallinity. Haemoglobin

The fact that these proteid substances can be artificially crystallized is not only interesting in itself, but is important as presumably furnishing a means for making preparations of undoubted purity, which will afford a sure basis for further studies of their properties.

Thomas Osborne, 1892[1]

Introduction

Crystals of calcite, quartz, and other minerals are among the delights of nature—fascinating to all observers since the beginning of time. And crystals have been studied by scientists since the earliest days, recording geometrical shapes, angles of inclination and the like with such enthusiasm that by 1911 the *Encyclopaedia Britannica* employed over 100 drawings to illustrate the different known crystal systems and symmetries for its article on 'crystallography'

Originally, studies were carried out with naturally occurring crystals of mainly inorganic substances—the French founder of the science of crystallography, René-Just Haüy (1743–1822), was in fact a mineralogist. However, scientists soon sought to produce crystals in the laboratory as a kind of self-purification, molecules coming out of solution by being selected on the basis of their ability to fit into the lattice of the growing crystal. Haüy took the principle of purity for granted: 'Ultimate particles forming a given mineral have the same shape. Different crystal habits reflect different ways in which identical molecules are put together.'[2]

Unfortunately, crystallinity was rarely found among plant or animal products. The term 'protoplasm', for example, the catchword for the as yet largely uncharacterized contents of living cells, was commonly defined as 'slimy and semi-fluid'. 'Gluten', the protein derived from wheat, one of the first proteins to be given a name, has its etymological roots in the word 'glue'. Gerritt Mulder (see Chapter 1), in reporting the procedures needed for purification of the proteins he analysed, described many of his products as 'syrupy' or 'frothy'.

Here is an explicit example, twenty years later than Mulder, but just as crude, namely the 1859 method by which 'myosin' was first isolated from frog muscles by Willie Kühne:

> Frog muscles were freed from blood by injection of 1% salt solution, then removed and frozen. After 3 hours the mass was cut up and pounded to a snow; on thawing it gave a syrupy liquid which was filtered through a linen cloth. A drop of this 'muscle plasma', falling into water at 0° gave a white, opaque precipitate which would redissolve in salt solution.... The liquid could be kept many days at 0° but on being brought into a warm room it quickly coagulated.[3]

No wonder 'real' chemists sneered at biochemistry, called it 'Schmier-chemie', which they continued to do until much more recently than 1859![4]

It must thus have been a godsend when that most easily accessible of animal fluids, blood, yielded crystals of the protein now known as haemoglobin and did so spontaneously.

Crystals of haemoglobins from many species

Crystal formation was first discovered accidentally in 1840, in samples of earthworm blood spread out on glass surfaces,[5] but crystallization could soon be induced deliberately, from the blood of virtually all animal species, by adding water to lyse the red blood cells and to solubilize their contents, followed by slight acidification with carbon dioxide. By 1871, the versatile physiologist Wilhelm Preyer (1841–1897) reported crystals for nearly 50 species from his own work and elsewhere—all manner of mammals (including man, of course), birds, reptiles, fish.[6] A generation later, in 1909, a truly remarkable book appeared, the result of more than five years' work by E. T. Reichert, a Pennsylvania professor of physiology, collaborating with his colleague A. P. Brown, professor of mineralogy, and sponsored by the Carnegie Institution of Washington. They extended the list of species several-fold—a mind-boggling zoo including hippo-potamus and elk, rodents galore, baboons of every variety, dogs and cats, moles, and bats, etc. etc. There are 600 glossy photographs at the end of the book, showing only a selection of the pictures of individual crystals

Fig. 2.1 Haemoglobin crystals, photographed in 1909 (from ref. 7). No. 69 is from the quail, no. 70 is from the guinea-fowl. Several hundred species were compared in all.

that they had available; crystal chracterization and precise measurements of angles of inclination of the crystal planes are reported in each case.[7]

Until the inception of this study it had not yet been conclusively determined whether corresponding proteins of different species of animals or of plants are chemically identical. Reichert and Brown provided the proof: haemoglobins from different species are certainly not identical. They hypothesized that:

> the observed differences in crystal form can serve as a useful index (gross though it be, with our present very limited knowledge) of those physico-chemical properties which serve directly or indirectly to differentiate genera, species, and individuals. In other words, vital peculiarities may be resolved to a physico-chemical basis.

No comparable effort had at that time ever been expended in investigating species differences at the molecular level.

Whether it was worth the effort is questionable. Crystal form is an overall 'morphology' only remotely related to molecular details and the hoped-for systematic relationship to biological classification or to evolution did not materialize. That sort of insight would have to await the more sophisticated approaches of the twentieth century, such as determination of amino acid sequences for corresponding proteins. On the other hand, the effect on the broad view must have been considerable. Protein science had progressed to the opposite extreme from Mulder's single universal protein or Liebig's four, for Reichert and Brown and their crystallography-minded predecessors conjured up vistas of a virtual infinity of proteins—thousands of unique species modifications superimposed on an ever-increasing variety of chemical and functional distinctions.

Crystallinity as criterion for purity

This abundance of haemoglobin crystals and the other proteins that soon became available in crystalline form sparked a lively debate on the question of how safely one could assume, in the explicit case of proteins, that crystallinity implied chemical purity. Doubts were expressed by Albrecht Kossel, for example, based on the knowledge that protein crystals can often absorb water and many foreign substances without loss of crystallinity.[8] And the topic of isomorphism, non-identical molecules giving rise to crystals of similar form, was of course repeatedly raised. Many of the later doubters—after around 1910—were driven by a prior agenda, avid believers in the colloid theory of formation of large particles (see Chapter 4), which did not accept that molecular reality or the concept of 'purity' extended to proteins at all. By having an agenda, they were able to put the burden of proof on those who argued against the agenda, and the absolute certainty they demanded was of course not realistically attainable.

Most people freely admitted that non-identical molecules can and do indeed sometimes crystallize together, but argued that there must be substantial similarity—some matching of shape and directional chemical affinity. A typical statement, picked more or less at random from published papers at the time, is that 'crystallizability makes it easier to get a pure substance than with an amorphous body'. Thomas Osborne, another authoritative voice of the day (as much so as Kossel), asserted that the fact that proteins can be crystallized is

> not only interesting in itself, but is important as presumably furnishing a means for making preparations of undoubted purity, which will afford a sure basis for further studies of their properties.[10]

In the case of haemoglobin, just looking at the photographs of the crystals, with their sharp edges and strictly defined angles, was probably fairly convincing. Does it seem likely that a motley mixture of molecules can create such a thing?

The question has quite recently been reopened in a historical context by N. W. Pirie,[11] who stressed that the answer must ultimately depend on what a protein preparation is intended for. If one is measuring enzyme activity, for example, then integrity of the active site on each molecule is what is important, leading to a perspective quite different from that held by chemists of an earlier day who could view identity of molecular shape and size as a satisfactory objective.

Nineteenth-century chemistry and spectroscopy of haemoglobin[12]

With all the attractions of easy access, biological importance, striking colour, and the very glamour of dealing with a crystalline material, it is not surprising that research on haemoglobin prospered. Both in quality and in quantity, work on this protein rapidly eclipsed anything done with other proteins.

Already before real molecular studies began, it was known that blood contains vastly more oxygen than could possibly be held in simple solution and that most of it was in fact bound to the colouring matter found in red cells, the protein first crystallized in 1840 and at that point not yet named. It was found chemically to be a conjugated protein, formed by association between a porphyrin-derived prosthetic group and the protein portion. The prosthetic group was identified as the source of the red colour in 1853 and was separately crystallized, a process that was quickly adopted by the courts as the basis for a forensic test for blood. The presence of iron was established, bound to the prosthetic group, which was named 'haemin'.*

The colourless protein portion was called 'globin' and the combined native molecule was named 'haemoglobin' in 1864 by Felix Hoppe-Seyler (1825–1895).[14] A huge amount of organic chemical research followed, to establish the chemical formula of haematin and the nature of its linkage to protein.

Especially remarkable are the early spectroscopic studies of haemo-

---

* The discovery that haemoglobin contained iron was made in 1853 by Ludwig Teichmann (1823–1895), a Polish anatomist.[13] The forensic test was Teichmann's chief claim to fame and for some 10 years it was the only recognized test for suspected blood stains on clothes, furniture, etc. Teichmann's main interest as a mature scientist was in the lymphatic vessels: he maintained that there was no connection between them and the blood capillaries.

globin, carried out almost immediately after the first chemical application of spectroscopy by Kirchoff and Bunsen in Heidelberg in 1860. The pioneer in these studies was the eminent German physiologist Felix Hoppe-Seyler[15] but unexpected personalities are also found among the researchers who became attracted to this protein, scientists outside the fields of chemistry or physiology, whose main interests would have been far from protein science. G. G. Stokes, for instance, who was Lucasian professor of mathematics at Cambridge, the post created for Isaac Newton and currently held by Stephen Hawking. Stokes, best known for his theoretical work on the motion of particles (Stokes' law) and on theoretical spectroscopy, was intrigued by the spectral differences between oxidized and reduced haemoglobin and made experimental measurements of them. He compared the results with the spectral difference between arterial and venous blood, and found them similar. He suggested that venous blood must be oxidized during its time in the lungs, even though that time is brief.[16] (This was in 1864. 'Biophysics' is not as modern an invention as some like to think!)*

Much of the work that followed after Stokes was lucidly summarized in an 1898 textbook by Arthur Gamgee, who was himself a spectroscopist. He describes the discovery of the ultraviolet Soret absorption band at a wavelength near 414 nm (named after the Swiss discoverer J. L. Soret), more powerful in its absorption intensity than the red band in the visible spectrum. He showed that spectroscopy can distinguish unambiguously between the various forms of haemoglobin: the reduced form; the oxygenated form containing reversibly bound oxygen, and its analogues with bound CO; 'methaemoglobin' in which the Fe atom of the haem group is irreversibly oxidized to the ferric state—and that the differences are similar regardless of the species origin of the protein. The equipment used to measure the spectra is illustrated in Gamgee's article and is seen to be remarkably advanced; photographs are provided of the spectra themselves and they are shown superimposed on the lines of the solar Fraunhofer spectrum, which provided accurate wavelength calibration.

---

*The tendency for work on haemoglobin to attract physical scientists never diminished with the passage of time. Seventy years after Stokes, Linus Pauling first entered the ranks of biophysicists by studying the magnetic susceptibility of haemoglobin with and without bound oxygen and used the results to define the bonding of the iron atom to the protein.[17] (This was 15 years before the α-helix.)

The sophistication is truly astonishing, especially when we appreciate that Gamgee's account is textbook material, the dates of original references covering two preceding decades.[18]

To this chemical/physical multiplicity of information, we must add the physiological aspect: molecular binding equilibria for oxygen and the other haem ligands, and parallel studies of whole blood. Barcroft's definitive physiological treatise on respiration contains much chemical detail about the reactions of haemoglobin with O_2, CO, etc.; it is not just about blood as a physiological fluid.[19]

Nothing like this wealth of knowledge exists from that period for any other protein. Compare myosin, for example, the protein identified from muscle about the same time as haemoglobin research began, and certainly as abundant and readily available as haemoglobin. But *not crystalline*: we have earlier described the crude and messy method for its isolation. It is, however, not this method that provides the most striking contrast in the present context, but the ignorance of what it was that had been isolated. It was to be almost 100 years before even the most minimal molecular information about myosin and its related proteins became available. Until then no progress of any kind could be made in understanding the mechanism of muscle contraction, as our chapter on that subject later in this book will demonstrate.

Other crystalline proteins

Seed proteins often existed in crystalline form within their native seed cells and their purification in crystalline form was first reported in 1855. In 1892 they became for many years the mainstay of the pioneering research of Thomas Osborne at the Connecticut Agricultural Research Station in America,[20] whose work we already quoted earlier in this chapter and who will be seen in Chapter 4 to have been a central figure in the advance of protein chemistry in general.

As protein research expanded, other crystalline animal proteins became known. They included the old protein war horse, egg albumin, which was first crystallized by Franz Hofmeister in 1889;[21] an improved method of preparation followed from F. G. Hopkins a decade later.[22,23] Hopkins clearly attached enormous value to being able to demonstrate that crystallinity could in this case be equated with purity: he characterized his product as a 'crystalline albumin, which, on repeated fractional crystallization, shows absolute constancy of rotatory power ($-30.7°$) and a constant proportion of sulphur . . .'. Serum albumin and milk albumin were crystallized about the same time with equal care. None of these proteins had any specific biological functions that might have provided a biochemical criterion for purity, which was part of the reason for being compulsive about physical data. It was not till much later (in 1926) that the first enzyme was crystallized by J. B. Sumner—another plant protein, urease from the jack bean.

Concluding comment

We shall see in subsequent chapters that haemoglobin has always remained at the forefront of protein research and eventually, in the modern era, its crystallinity became itself a crucial factor, when X-ray diffraction was developed as a tool for the determination of high-resolution three-dimensional molecular structures. Quite appropriately, haemoglobin and its close relative, myoglobin, were the first proteins to which this technique was successfully applied. This revolutionary application of crystallinity could not have been anticipated in 1850 or 1900, even by the wildest imaginations. Röntgen only discovered X-rays in 1896, and it was much later before the potential value of X-ray diffraction patterns from regularly ordered arrays could be appreciated.

CHAPTER 3

The peptide bond

The type of condensation described here through formation of
–CO–NH–CH= groups may thus explain both the *building up* of
protein substances in the organism, as well as their *breakdown in the
intestinal tract* and in the tissues. On the basis of these given facts one
may therefore consider the proteins as for the most part arising by
condensation of α-amino acids, whereby the linkage through the
group –CO–NH–CH= has to be regarded as the regularly recurring
one.

Franz Hofmeister, 1902[1]

Introduction: proteins are built up from amino acids

By 1900 it was understood that protein molecules are mainly built up
from amino acids and most of the constituent amino acids had been
identified—a huge change in the frame of reference compared to 1840,
the year of Mulder's elemental analyses, when only glycine and leucine
had been known. The rest had gradually appeared over the years, isolated
from protein hydrolysates and chemically characterized. Emil Fischer,
one of the two principal figures in this chapter, himself added two to the
list, namely proline and valine, first isolated in his laboratory in 1901.
One could not know at the time what the ultimate total would be, but in
fact only three remained to be discovered—five if we count asparagine
and glutamine. Actual dates are given in Table 3.1.

Retrospectively, asparagine and glutamine merit special mention. We
now realize the similarity between their amide bonds and the peptide
bonds of proteins: asparagine and glutamine would normally have been
hydrolysed to the corresponding acids in the course of protein breakdown. This hydrolysis was actually anticipated as early as 1873 by
Hlasiwetz and Habermann, on the basis of the knowledge that ammonia
is always a product of complete protein decomposition.[6] They considered it very likely that the ammonia was originally derived from

Table 3.1 Dates of recognition of amino acids as protein constituents, based on isolation from protein hydrolysates

1819	Leucine	1889	Lysine
1820	Glycine	1890	Cystine/Cysteine
1846	Tyrosine	1895	Arginine
1865	Serine	1896	Histidine
1866	Glutamic acid	1901	Valine
1869	Aspartic acid	1901	Proline
(1873	Asparagine, see text)	1901	Tryptophan
(1873	Glutamine, see text)	1903	Isoleucine
1875	Alanine	1922	Methionine
1881	Phenylalanine	1936	Threonine

The dates are based on reviews by Fischer (1906),[2] Cohn (1925),[3] Vickery and Schmidt (1931),[4] and Vickery (1972).[5]

asparagine and glutamine residues that emerge afterwards as aspartic or glutamic acid. The idea was probably accepted by many people and in Table 3 we have given 1873 as the date of first 'recognition' of the two amidated amino acids as real individual protein constituents. Purists, of course, set the date at 1932, when organic chemists first learned how to break peptide links while keeping side-chain amide bonds intact, and asparagine and glutamine could thereby be unambiguously established as originating from a native protein.[5,7]

This productivity in the enumeration of amino acids must be viewed against fundamental changes in the general chemical background that were going on at the same time. In particular, the concept of 'bonds' linking atoms into molecules had become established.[8] Thus the manner in which the amino acids derived from protein decomposition were originally linked in the native state became a natural concern for organic chemists. On the other hand, vital individual distinctions, like ionic charges or affinity for water, would not yet have been uppermost in anybody's mind. Amino acids were always given their unadorned organic chemical formulas: $H_2N-CHR-COOH$, with 'R' variable, and with all the atoms electrically neutral.*

---

*As Franz Hofmeister pointed out in the quoted passage at the head of this chapter, the amino acids in proteins are all α-amino acids, meaning that the amino nitrogen atom is always attached to the same carbon atom as the acidic COOH

Germany in the forefront

It was not only the world of protein chemistry that had been transformed since the time of Liebig and Mulder; the organizational infrastructure within which science operated had also altered dramatically. Germany was on top of the world. They had won the Franco-Prussian War in 1871 and had annexed Alsace; Strasbourg became Strassburg and the university there (where Louis Pasteur had begun his professional career in 1849) was made into a showcase for German superiority. German influence extended eastwards as well: Prague (more accurately 'Prag') was a German city, its university called the 'Deutsche Universität'.[9]

With German dominance, huge research groups became the order of the day, presided over by powerful directors, following the precedent set 50 years earlier by Justus von Liebig in Giessen. They flourished everywhere and communication between them was greatly improved by the existence of railway networks to link them and by nationwide meetings, where issues of the day could be debated. A prominent example is provided by the meetings of the Gesellschaft deutscher Naturforscher und Ärzte,† an organization that had gained increasing importance since its modest origins in 1822[10,11]—its very name suggests an invitation to biologically oriented scientists, who in other countries might not be welcomed at gatherings that tended to define 'science' more narrowly.

It was at the 74th meeting of this society in 1902, in Karlsbad (now in the Czech Republic), that the probable solution to the problem of bonding between amino acids was announced—independently by Emil Fischer and Franz Hofmeister.[12] It was quite a coincidence to have the two presentations on the same day at the same meeting (Hofmeister in

---

group. Moreover, all but glycine are optically active, invariably in what is known as the L-configuration (the D-configuration being the mirror image). It needs to be appreciated that the relationship between molecular asymmetry and optical rotation would have been common knowledge at the time, already well understood for more than a generation, ever since the publication of van't Hoff's famous pamphlet on the tetrahedral carbon atom in 1874. Understanding of this quite sophisticated aspect of organic chemistry was far advanced when compared to knowledge of the structure of proteins.

† Society of German Scientists and Physicians. The literal translation of 'Naturforscher' is 'explorer of nature'.

the morning, Fischer in the afternoon), and the validity of the conclusion was enormously strengthened by the fact that the two presenters used entirely different styles in their approaches to the problem. It was a watershed, a rare instance where one can point to a single day on which, in effect, the history of an entire field was changed. [The solution of the linkage problem has been described by Vickery and Osborne as 'probably the most important event in the whole history of protein chemistry'.[13] From their vantage point (1928) that would seem a justified verdict.]

Emil Fischer (1852–1919)[14,15]

Fischer and Hofmeister were almost exact contemporaries, but differed in most other important respects—training, research specialization, manner of interaction with students and assistants. Fischer was the more famous of the two, the personification of an organic chemist—rigorously pursuing synthesis and proof of structure. Wilhelm Ostwald, the founder of physical chemistry, described encountering him at a meeting in 1889:

> I found myself in a swarm of organic chemists gathered about Emil
> Fischer, already regarded as the future leader of our science, since what was
> not organic chemistry was not recognized as chemistry. [16,17]

Fischer had been one of those who profited from Germany's successes on the battlefield, transferring in 1872 to the University of Strasbourg (we shall use the restored French spelling). There he obtained his Ph.D. under the direction of Adolf von Baeyer, one of the most creative organic chemists of all time.*

Fischer accompanied von Baeyer to Munich when the latter succeeded Liebig there in 1875, then held successive faculty appointments at Erlangen, Würzburg, and Berlin.[18] He had been professor at Würzburg, already famous, at the time of the encounter with Ostwald. In Berlin he com-

---

* Von Baeyer was one of the last of the breed of devoted synthesizers, who stayed at the laboratory bench all day alongside their students and assistants, leading them by example as well as by instruction. Fischer emulated Baeyer's personal involvement at the bench and suffered for it. He had mercury poisoning in 1881, resulting from formation of gaseous $HgEt_2$ in one of his reactions. In 1891 he was seriously incapacitated by the insidious long-term effect of phenylhydrazine vapour—phenylhydrazine was a central reagent for his synthetic activities in the sugar field.

Fig. 3.1 Emil Fischer. (Source: Photo Deutsches Museum München.)

manded an institute for 250 people, the largest in the world. His earliest and probably best work (from the standpoint of organic chemistry) was on carbohydrates and purines, for which he received the Nobel Prize in 1902. However, his work in that field included the study of enzymatic cleavage of glycosides, which aroused an intense interest in enzyme specificity and led to Fischer's famous 'lock-and-key' hypothesis, which we shall mention in the context of enzyme function in Chapter 15.[19] This aspect of his work also led directly to Fischer's involvement with proteins. Although the question of whether enzymes are protein molecules was still being debated then (and still more so even later), Fischer was a believer, as shown by the following quotation from a letter written in 1905 to von Baeyer:

> My entire yearning is directed toward the first synthetic enzyme. If its preparation falls into my lap with the synthesis of a natural protein material, I will consider my mission fulfilled.[20]

Once his enthusiasm for applying his skills in organic synthesis to polymerization of amino acids had become aroused, all the resources of Fischer's laboratory were turned in that direction, with the explicit hope even then that they would lead the way to 'the manner in which proteins may ultimately be synthesized'.[21,22] Fischer invented the terms 'peptide' and 'polypeptide'; he improved laboratory methods for analysis of amino acid mixtures derived from cleavage of proteins or from his synthetic analogues; he discovered two new amino acids in the process. His first synthetic polymeric product was glycylglycine and by the time of the Karlsbad meeting he had produced at least tetrapeptides. And the synthetic products gave characteristic colour tests and were degraded back to amino acids under the same conditions as were used to hydrolyse proteins extracted from living organisms.

The text of Fischer's lecture at Karlsbad was not published in full, but only as an 'Autoreferat', an abstract written by the author himself after the presentation.[23] He therefore was aware, at the time of writing, of Hofmeister's earlier lecture, about which he had not known in advance. In the abstract, he appropriately defers to Hofmeister's priority in the conviction that peptide groups are responsible for the coupling of amino acids in protein molecules, but points out that that was an inherent assumption of his work on polypeptides. The question of priority has never been seriously argued, nor should it be. In the long run, Fischer's rigorous chemical methods probably carried the greater weight. No organic chemist worth his salt believes in a chemical structure until he can synthesize it with his own hands!

Franz Hofmeister (1850–1922)[24,25]

Franz Hofmeister was a physiologist, chemically inclined in the way he thought about the subject, but never a chemist in practice in any sense that Fischer would have recognized—*ab initio* syntheses did not figure in his projects. He was the son of a prosperous Prague physician and became involved with proteins from the beginning of his career—his mentor at the university in Prague (Hugo Huppert) was a disciple of the

Fig. 3.2 Franz Hofmeister. (Source: Science Museum/Science & Society Picture Library.)

physiologist Carl Lehmann (1812–1863), the acknowledged expert on nutritional degradation of proteins in the intestines. This degradation requires several successive enzymes and 'peptone' and 'albumose' were the best known among many terms used to designate intermediates between the original macromolecule and the ultimate amino acids produced. These intermediates were at the time given more significance than they merit because it was thought that they were the *final* product of events in the digestive tract, able to cross the intestinal wall for synthesis of new proteins without the need for all-the-way hydrolysis. Hofmeister's *Habilitationsschrift* in 1879 dealt with analysis of such peptones and undoubtedly the question of their structure would have been an enduring legacy.

Hofmeister, unlike Fischer, was somewhat of a recluse and politically uninvolved. (No jockeying for positions in his career.) He continued to reside in his old family home after he became professor in Prague in 1885, even after he married. When he moved to Strasbourg in 1896, he became just as entrenched there; he had offers of more prestigious posts, but turned them down. While at Prague, Hofmeister became interested in protein purification, with emphasis on the ability of inorganic salts to precipitate proteins from solution—the process of 'salting out'. This led to the discovery of the famous 'Hofmeister series', the order of effectiveness of the salts he used. Largely on the basis of his results, ammonium sulphate became the protein chemist's favourite precipitation agent, used for decades afterwards in the purification of the endless number of enzymes that emerged from biochemical laboratories. Hofmeister himself used ammonium sulphate in an effective method for producing crystalline egg albumin.[26] Most importantly, he was a steadfast and influential advocate of the importance of enzymes—living cells are mostly bags of enzymes, if we can be permitted to oversimplify. In Strasbourg, both before and after the Karlsbad meeting, enzyme action became the dominant theme in Hofmeister's laboratory.[27]

Hofmeister's lecture at the 1902 Karlsberg meeting is much better documented than Fischer's. A semi-popular version of it was published,[28] based on a detailed article published simultaneously.[29] (The *definitive* paper from Fischer did not come until 1906.[30]) These articles show Hofmeister's approach to the problem of intra-protein bonding to have been entirely deductive, not based on purposeful experiments of his own. He argued against several conceivable bonds, such as C–C bonds or ether or ester links, because their breakdown by enzymes such as trypsin would be difficult to comprehend—for example, neither long nor short hydrocarbon chains are affected by trypsin. Linkage through nitrogen alone ($=$C–N–C$=$) was ruled out in addition because it would leave the protein with a huge number of free –COOH groups and a corresponding strongly acid character, which is not in fact observed.

On the positive side, Hofmeister attached particular importance to the 'biuret reaction', a colour reaction induced by copper sulphate, which organic chemists had known for a long time to be specific for sequentially linked amide groups. (Biuret itself has the formula NH_2–CO–NH–CO–NH_2.) A positive biuret test is obtained for all proteins and for hydrolysis intermediates such as the peptones mentioned above, but not by free amino acids. The simplest interpretation is that amino acids in proteins are linked by what eventually became known as 'peptide bonds'. To quote Hofmeister directly:

> On the basis of these given facts one may therefore consider the proteins as for the most part arising by condensation of α-amino acids, whereby the linkage through the group –CO–NH–CH$=$ has to be regarded as the regularly recurring one.

Aftermath: dreams unrealized

Fischer was single-minded in his scientific pursuits, indifferent to literature, music, art, or other cultural distractions. Furthermore, he ruled his laboratory in an autocratic fashion, expecting to be able to direct all its resources towards a single goal. And this is what he did with respect to polypeptide synthesis. By 1906 about 65 peptides of different chain length and amino acid composition had been made; eventually there were more than a hundred, the longest being an octadecapeptide made up of 15 glycines and 3 leucines.[31] But Fischer soon became disenchanted and he knew that his dream of 'the first synthetic enzyme' would never be realized. It would require the synthesis of polypeptides with amino acid sequences that would match those obtained from hydrolytic cleavage of enzymes, ultimately resynthesizing an enzyme from purely synthetic precursors. His synthetic methods were too primitive for that, both cumbersome and costly; for example, the amino group at what was intended to be the terminal member of a chain always had to be blocked with a substituent that could be selectively removed at the end of the synthesis.

Such methods, moreover, were not readily adapted for incorporation of active groups that required additional protection—for example, the side chains of lysine and glutamic acid.

Besides, when one thinks of matching protein cleavage products, the number of possible sequences becomes unmanageable. After 1910, Fischer's work on peptide synthesis ceased; his laboratory returned to carbohydrate and purine chemistry and he began a venture into nucleotide synthesis.

Hofmeister's laboratory was much less tightly focused than Fischer's: the work on enzymes in his laboratory had never represented a concerted group attack on particular problems along lines determined by the leader. Hofmeister, rather than assigning problems, suggested them, gave advice, and helped prepare papers for publication even though his name would not be on them. As a consequence, the work in his laboratory had always been characterized by diversity in experimental approach and this continued to be true. Since Hofmeister's name did not normally appear on publications from his laboratory, it becomes relatively difficult to trace patterns in what may have been going on at the laboratory bench, but chemical bonding had never been an experimental topic for him and this remained true after the Karlsbad meeting.

However, Hofmeister's acute awareness of the details of chemical procedures stands out and he was fully sympathetic to the problems that caused Fischer to retreat. In 1908, for example, he commented on analysis of hydrolytic products:[32]

> The enzymatic breakdown of proteins always leads to mixtures of many substances of unequal molecular size . . . and the isolation of definite chemically-characterized albumoses or peptones is a thankless task

and just as forcibly on synthesis:

> The rapid progress made in the synthesis of polypeptides under the aegis of Fischer could lead to the view that one may expect the elucidation of protein structure to come entirely from this approach. This hope at once comes to naught if one considers the enormous number of synthetic possibilities.

Knowledge of how one amino acid is chained to its neighbours evidently did not enhance the immediate prospects for protein synthesis in the laboratory. Confidence in the peptide bond made no difference. It did not indicate a liberating pathway to the ultimate goal; progress would require decades of ingenuity and hard labour.

A few years later, of course, progress of any kind was effectively halted by World War I, a war instigated, it is said, by inflated visions of glory in the mind of the German Kaiser. When Germany lost, it meant the end of its national glory. Strassburg became Strasbourg again, Prag was re-named Praha and became the capital city of Czechoslovakia. All scientists were affected, even those who might have seemed remote from military ambitions. (Surely protein chemists would belong to that number?)

Fischer in particular was tragically affected—more unrealized dreams. As we said earlier, he had derived benefit from the victory of 1872, and in 1914 he was at first an ardent patriot. He believed in the justice of Germany's cause and denied reported atrocities on conquered popula-tions. On the professional level, he undertook responsibility for organizing the production of chemicals needed for the war effort. But he eventually became disillusioned and in January 1918 he was co-signatory of a mem-orandum sent to heads of military and civil governments urging an end to the war, explaining that science and technology were helpless to meet the needs of maintaining the war. The warning was not heeded.

Fischer became depressed, especially after two of his sons had been killed in battle. His health remained poor, too, with the effects of phenyl-hydrazine never completely overcome. Fischer committed suicide in 1919.*

The level of Hofmeister's support for the war is not well documented. He was, of course, forced to leave Strasbourg, but his Czech origins en-titled him to Czech citizenship and a possible position in Prague. He declined that opportunity and preferred to remain with his former German colleagues. He took a position at the University of Würzburg and he died in that city in 1922.

---

* Fischer's oldest son, Hermann Otto, survived the war, joined the Banting Institute in Toronto, and eventually became professor of biochemistry at the University of California. He brought with him from Germany a collection of 9000 reference compounds made and bottled by his father. Immunologist E. A. Kabat in the 1950s tested some of the peptides in this collection by paper chromatography and in all but one found only a single spot—a test for purity that had, of course, not been available to Fischer himself.

Challenges to the peptide bond

The simultaneity of the Fischer and Hofmeister papers might lead one to suspect that the peptide bond was 'in the air'—a common idea in many minds—and that Fischer and Hofmeister just happened to be first to articulate it for public presentation. The evidence is against that view and indicates that few people had yet entertained the idea of a unique or predominant bond. Albrecht Kossel, for example, in many ways one of the most far-seeing of protein scientists at the time, was still under the influence of the then current ideas about protein digestion in the gut, which we have already mentioned as an early influence on Hofmeister. It was thought that differently constituted parts (peptones, albuminoses) were the *final* product of events in the digestive tract, able to cross the intestinal wall for synthesis of new proteins without the need for all-the-way hydrolysis. A logical corollary was that native proteins were originally built up from such fragments, with local atomic arrangements maintained intact.[33]

Kossel's major influence on biochemistry came through his pioneering work on the constituents of the cell nucleus (Nobel Prize in 1910), which included the protamines, primitive substances related to proteins, but less complex, arginine being by far the most abundant constituent amino acid. Kossel made the plausible suggestion that a simple arginine-rich 'protamine nucleus' was the core of protein structure, to which other amino acids and larger structural units could be attached. This was in a lecture delivered to the German chemical society in June 1901, that is, just a year before the Karlsbad meeting. There is no inkling in the presentation of any conception of a universal chaining mechanism, though Kossel was quickly converted to the polypeptide theory thereafter.[34]

Many other alternatives persisted for some time; they are discussed in a scholarly review by Vickery and Osborne.[35] One of the unconverted, even long after the Karlsbad meeting, was Emil Abderhalden, who was for some years Fischer's closest and most valued associate, prized by him for his synthetic skills in the laboratory. Abderhalden (1877–1950), a Swiss national, became the most insistent challenger—persistent as well, for his challenge was at its height in 1924, long after the polypeptide theory had become textbook material.[36]

Diketopiperazines, cyclic double anhydrides of two amino acids,

$$
\begin{array}{ccc}
\text{NH} & - \text{CHR} - & \text{CO} \\
| & & | \\
\text{CO} & - \text{CHR} - & \text{NH}
\end{array}
$$

were at the heart of Abderhalden's theory, with subsequent complexity arising from 'alternative junctions' through the R groups.[37,38] There was a certain logic to the idea because the pathway of Fischer and Fourneau's original synthesis of glycylglycine had in fact been via diketopiperazine—the double anhydride came first and was converted to the dipeptide by what the authors called 'arrested hydrolysis'. This shows how primitive the knowledge of peptide chemistry had been in 1901, but to go back to all this in 1924 was not very sensible.

As Vickery and Osborne commented in 1928, the synthesis of diketopiperazines may be easy, but the difficulty is to perceive any connection between them and proteins; no enzyme found in nature has ever been found to split a synthetic diketopiperazine ring! Their final comment was 'Hypotheses that disregard well-established facts of protein chemistry do not tend to promote progress'. (Abderhalden seemed throughout his career to relish controversy, almost inevitably misjudging which would be the winning side.)

And even later, in 1936, there was the 'cyclol' hypothesis of Dorothy Wrinch, a bizarre geometrical construct in which the peptide link was rejected in favour of an arrangement of six-membered rings,[39] supposedly to account for the remarkably compact overall shape that many protein molecules were by then known to have (see Chapter 10). The theory would not be worth mentioning here (there are always a few crazy ideas floating around) were it not for the fact that Wrinch was taken seriously, given a platform at international meetings, a paragraph or two in scholarly textbooks. It is not easy to understand the reason for this. Whereas Abderhalden was a former associate of the great man himself, always spoken of by Fischer with great praise, which could be considered to justify giving him an initial hearing, Wrinch was pure bravura, not even a chemist, and unable to provide any chemical evidence for her hypothetical structures.

CHAPTER 4

Proteins are true macromolecules[1]

Previous analyses of haemoglobin have led to the surprising result that this compound contains 600 atoms of carbon for each atom of iron, i.e. at least 600 atoms of carbon in one molecule. Haemoglobin is therefore incomparably more complicated than all chemical compounds so far known to us. However, the reported size is a minimal number. It was calculated by assuming that haemoglobin contains only one atom of iron.

From the foregoing numbers it follows without doubt that for 1 atom of iron haemoglobin contains exactly 2 atoms of sulphur and that haemoglobin is a chemical individual.

O. Zinoffsky, 1886[2]

Introduction: the colloid/macromolecule debate

Do macromolecules exist? Can amino acids be linked together to form longer and longer polypeptide chains, *ad infinitum*? Emil Fischer had given up on his synthetic efforts at the stage of an 18-mer: does that imply a limit? Debate on these questions was intense in the early part of the twentieth century, fuelled by colloid chemistry, a new way of thinking about large particles based primarily on the properties of *inorganic colloids* (colloidal gold, hydrated silica, etc.). These particular colloids were known beyond any doubt to exist as large particles, but mass or size was variable, depending on environment and conditions of formation; these particles were obviously *aggregates* of the true molecules of the intrinsic components. The question of how such aggregates were formed and stayed in solution and how large they could become intrigued many brilliant minds and led to the development of much elegant theory and powerful new analytical tools.

Some organic substances (rubber, cellulose, proteins) behaved in some respects like inorganic colloids. They, too, could reside in solution as huge particles and it was easy to suppose that they, too, were not molecular, but poorly defined clusters of much smaller 'true' molecules. And this supposition was enthusiastically supported by many of the most illustrious organic chemists of the day. They were used to working with hundreds of well-defined compounds of C, H, O, and N, which could readily be purified and analysed, and which invariably had molecular weights in the range of a few hundred at most. It seemed incredible and unreasonable to the organic chemists that substances of similar sorts of composition, readily soluble in water (proteins, polysaccharides) or benzene (rubber, polystyrene), should have molecular weights of several hundred thousand.

Thus the colloid/macromolecule debate was born, lasting till around 1930 before the last embers of dispute died down and macromolecular chemistry became established as an independent field of science. The conventional history of this debate, with almost the status of classical folklore, casts Hermann Staudinger (1881–1965) as the hero, the knight in shining armour, and colloid chemists as his principal adversaries. Staudinger is named as the *very first* effective proponent of the idea (1920) that true molecules of huge size are capable of stable existence.[3–8] The colloid chemists are viewed as unflinchingly opposed, brooking no resistance to the dogma that what might seem to be macromolecules are nothing more than loose aggregates of much smaller entities. The colloid chemists are held to have had a malign effect not only on their own field (as seen retrospectively) but also on the development of biological sciences; for example, Florkin and Stotz[9] refer to 'the dark age of biocolloidology'. In the end, of course, truth wins out. After a long, acrimonious battle in which even Martin Luther features ('Here I stand, I can do no other', quotes our hero)[10] Staudinger is vindicated by incontrovertible experimental results and suitably rewarded with the Nobel Prize for chemistry, though rather tardily, not until 1953, long after the controversy had ended.

This version of macromolecular history has become a part of many histories of chemistry as a whole and can even be considered accurate from the narrow point of view of 'polymer chemistry', as the term is normally used today, but it is simply not true if we include proteins within our perspective. Protein scientists paid little attention to the controversy[*]

---

[*] Enzymologists are often cited as exceptions. They did not dispute the large size of protein molecules, but did resist identification of enzyme activity with the protein. See Chapter 15.

and went merrily on their way, in no way stifled, as the following account will demonstrate.

Elemental analysis convincingly demonstrates the probable huge size of protein molecules

The basic reasons for protein science's virtual immunity from the colloid fracas are inherent in the first three chapters of this book.

1. Almost a hundred years before Staudinger, in the very first paper in which G. Mulder introduced the term 'protein' to designate albumin-like substances,[11] it was clearly recognized on the basis of elemental analysis that protein molecules are much larger than those of many common organic substances.

2. Shortly after Mulder, haemoglobin arrived on the scene, its crystals not only beautiful to behold, but also indicative of chemical purity, order, reproducibility. It elevated at least this one protein into the category of exact chemistry.

3. The discovery of the peptide bond provided the mechanism whereby long chains of amino acids could be created. Surely, this made the case for macromolecules all but airtight, no matter what others might be led to believe on the basis of experiences that excluded any contact with proteins.

In reference to elemental composition, haemoglobin was, of course, the star of the show. All analytical data, regardless of species or crystal form, demonstrated a fixed but extremely low content of iron, which led to a minimal molecular weight of around 16 000. There were arguments as to whether crystallinity can in general guarantee purity, which we discussed in Chapter 2. But in this case the crystallinity, coupled with the constancy of the iron content, convinced almost everyone. By 1872 it had become textbook material; for example, we can cite the manual of chemical physiology written by J. L. W. Thudicum (1829–1901), a former student of Liebig's, who practised medicine in London and was frequently called upon by the British government to serve as medical consultant:

> Persons who have not studied this branch of chemistry and ... do not read of atomic weights rising above 500, may wonder at the high atomic weight here assigned (to haemoglobin). But this body can now be obtained pure ... and always contains 0.4% iron.[12]

(Chemical nomenclature was still in a primitive state and 'atomic weight' was often still used to represent the sum of the weights of atoms in the molecule.)

The definitive paper on haemoglobin came in 1886 from Oscar Zinoffsky,[2] who worked in the laboratory of Gustav Bunge in Basel. He emphatically brushed aside the suggestion that had been made by some that haemoglobin is not a 'molecule' at all, but just a product of mechanical association of Fe-containing haem with a protein entity. He reinvestigated the Fe content and did analyses for elemental sulphur as well. Assuming 2 S per molecule, both measurements were compatible with the same minimal molecular weight of 16 700. Soon thereafter the high molecular weight was supported by the integral stoichiometry of the binding of oxygen or carbon dioxide to the 'molecule' that one had been led to define on the basis of the iron content—binding capacity was one ligand per 16 000 grams of protein.[13]

Around the same time amino acid content began to supplement elemental analysis, although the technology for anything like complete amino acid analysis was still a long way off. And with respect to amino acids the situation is the same as for elemental analysis: one or more of the amino acids are often in short supply in a protein hydrolysate, leading inevitably to a high minimal molecular weight.[14,15] Moreover, the same intrinsic difference between proteins and other natural products (cellulose, starch, rubber, all now known to be macromolecular) makes its impact again. The latter are all homopolymers and no inkling of large size could therefore be deduced for them in this manner. No wonder the received heritage of the group of organic chemists that laid the foundations of the synthetic polymer field was different from the received heritage of the investigators studying proteins!

One other approach to the subject merits mention. Hlasiwetz and Habermann,[16] two relatively obscure Czech chemists, have become quite highly regarded by historians for work done in1871 that amounts to the first serious attempt to account completely for the composition of a protein (casein) in terms of the products of its degradation by oxidation or hydrolysis. A serial procedure was used, from which successively simpler compounds emerged at each stage. The obvious conclusion was that the undegraded starting material must have had a very high molecular weight to account for the sequential appearance of new products; at the same time the methods of degradation endorsed the idea that 'ordinary' linkages were being broken at each stage. This basic idea that serial degradation provided proof for high molecular weight quickly became part and parcel of the broad consensus about the nature of proteins. It was, for example, explicitly stated in 1901 by Alfred Kossel,[17] without any reference to Hlasiwetz and Habermann's paper.

It is worth noting that Heinrich Hlasewitz (1825–1875) provides an interesting contrast to more famous figures (like Emil Fischer), having never been in the limelight, never leader of a huge research group. Hlasewitz was professor of chemistry at the University of Innsbruck and later at Vienna Polytechnic. His research dealt with natural materials of all kinds, with no special interest in proteins—in fact, he used the degradation principle as an early indication of the macromolecular nature of polysaccharides as well as for proteins. He loved music almost as much as chemistry and composed an opera as well as smaller works. L. Barth, one of Hlasiwetz's students, has written an affectionate (and recommended) biography.[18]

The rise of colloid chemistry

While organic chemists were making such good progress in understanding proteins, a new discipline—physical chemistry—was born,[19] which, among other virtues, promised new avenues to the measurement of molecular weight and dimensions. The year 1887 is often given as the year of coming of age, being the year when the *Zeitschrift für physikalische Chemie* was launched under the editorship of Wilhelm Ostwald, and an early application to proteins was reported soon thereafter: use of freezing point lowering to yield a molecular weight of 15 000 for ovalbumin.[20] This method, however, does not measure mass directly, but gets it indirectly from the number of molecules per gram of solute. The uncertainty is huge for macromolecules, where the number of molecules is necessarily tiny and its measurement easily distorted by error in the count of small molecules. The result was therefore quite appropriately treated with caution. However, no directly contradictory result was reported, and the 15 000 figure was widely quoted for its consistency with the more reliable weights based on elementary analysis. (Much later, of course, physical chemistry became a prime tool on its own for protein chemists, as it did for everyone else.)

Some years earlier, in 1861, the Scotsman Thomas Graham unwittingly laid the foundations of another new branch of science, namely colloid chemistry. It arrived initially with much less fanfare than physical chemistry, but proved ultimately to have a stronger impact on the orderly progress of molecular weight determination. We summarize here some highlights in its rise to popularity.

Thomas Graham (1805–1869), sometimes called the 'dean of English chemists' of his day, was a founding member and first president of the Chemical Society.[21] His best-known research had been on gaseous diffusion but in 1861 he turned his attention to liquids, measuring rates of free diffusion and of dialysis through various semi-permeable membranes (for example, parchment paper, gelatinous starch).[22] He found that some dissolved substances—hydrated silicic acid, hydrated alumina, starch, albumin, gelatin, etc.—had free diffusion rates that were many times slower than those of inorganic salts or sugar and were almost entirely blocked in the experiments using membranes. Graham coined the term 'colloid' to describe these compounds and the term 'crystalloid' to refer to inorganic and organic molecules that diffused rapidly. He chose the name 'colloid', based on the Greek word for glue, because gelatin (major protein component of natural glue) was supposedly a 'typical' example. Graham believed colloids to be large particles, probably *aggregates* of crystalloid molecules. Their supposedly *general* properties (in addition to low diffusibility) included absence of power to crystallize as aggregates and 'mutability', that is, a tendency to change in physical characteristics. The generality of these characteristics soon proved untenable, of course—the true believers would always have difficulty with haemoglobin crystals in particular. (In retrospect, the protein-linked name 'colloid' was perhaps a poor choice.)

Graham died in 1869, and for the next few decades academic interest in colloid chemistry was muted. August Kekulé was aware of his work and needs to be mentioned because he was the founder of the modern idea of specific bonds between atoms and his comments on the distinction between primary and secondary bonding, which would become one of the prominent features of colloid theory, are therefore of special importance. Kekulé, in a lecture[23] given in 1878, in fact gave short shrift to the idea (called 'peculiar aggregation' by Graham): is not the polyvalency of atoms such as carbon sufficient to explain the formation of huge net-like or sponge-like masses? Wilhelm Ostwald[24] barely mentioned colloids in his 1884 textbook of physical chemistry, but was more sympathetic than

Kekulé to the notion of 'peculiar aggregation': he expressed the opinion that colloidal solutions 'are rather mechanical mixtures than compounds'. The small number of people who were directly engaged in colloid research paid little attention to proteins—inorganic colloids tended to be easier to procure and work with. For example, the Dutch chemist van Bemmelen, a relatively prolific worker on colloids, does not mention proteins at all in a well-regarded 1888 paper focused on *hydration*[25]—a phenomenon that should surely make one think of the many highly water-soluble proteins if one ever thought about proteins at all.

Two British colloid chemists, Harold Picton and Stephen Linder, did include proteins in their research and seem to have had no difficulty in accepting their high intrinsic molecular weights. Picton and Linder are celebrated for their pioneering work (1892) on the movement of colloidal particles under the influence of an electric field.[26] They investigated substances like colloidal arsenious sulphide, which were clearly secondary aggregates, but they also included crystalline haemoglobin; they had no hesitation in accepting the molecular weight derived from elemental analysis and, in fact, pointed out that it must be a minimal value. In more general terms, they felt compelled to deny the sharp division between colloids and crystalloids and to advocate instead a continuity between crystalloids of high molecular weight in true solution and the more traditional kind of colloidal suspension. Colloid chemistry might have quite naturally evolved in a much less strident and controversial direction[27] had it not been for the missionary zeal of Wolfgang Ostwald (son of the aforementioned founder of physical chemistry) who entered the field a decade later.

Some chemists who did not themselves specialize in the colloid field became enthusiastic advocates. A potent voice was that of A. A. Noyes, one of the most important figures in the growth of American chemistry,[28] often dubbed 'King Arthur'. Noyes was elected the youngest ever president of the American Chemical Society in 1904 and, though he himself had no previous involvement with colloids, chose to deliver an extraordinary inaugural address[29] to express enthusiasm for the field, complete with experimental demonstrations from the lecture podium. He followed the conventional definitions, calling colloids 'this important state of aggregation' and justified his review of the subject because of existing contradictions in the literature. In retrospect he was (being a novice) quite prescient, for he felt it necessary to define two different classes of colloidal mixtures: colloidal solutions (as typified by gelatin and ovalbumin) and colloidal suspensions (as typified by arsenious sulphide), differing in viscosity, tendency to coagulate, and other properties. Noyes, however, offered no theoretical insights to account for this difference. And the possibility that the idea of 'colloid' as a unifying concept might be artificial does not seem to have occurred to him at all. (Might he not be dealing with two quite different classes of chemical substances, sharing only the property of large particle size?) On the contrary, Noyes explicitly reiterated the concept that 'proteins consist of aggregates of the ultimate chemical molecules'. Given his enormous prestige, he must have exerted a potent influence on at least part of the chemical community—he himself did not subsequently return to colloids in the laboratory or on the podium.

One might mention in passing that the universal and ageless dream of finding a simple key to the secrets of life gained some adherents for colloid chemistry and kept it in the news. Graham planted the seed in his defining paper back in 1861, when he speculated that the 'peculiar physical aggregation' of colloids might be a necessary ingredient of 'substances that can intervene in the organic process of life'. This concept remained popular for a long time—though not usually, it should be said, among scientists who had experience of their own with the chemical substances (proteins, enzymes, etc.) that actually participate in life processes. Some adventurous souls had visions of magic medical benefits. The Frenchman A. Lumière is an example,[30] claiming that 'l'état colloidal conditionne la vie; la flocculation détermine la maladie et la mort'. In America, Martin Fischer, one-time protegé of Jacques Loeb, became a celebrated 'colloid healer', and is said to have become rich by practising his art on wealthy patients.[31]

But the most notorious and bizarre example of this sort of thing came from a highly placed professional physical chemist—Wilder Bancroft (1867–1953), professor at Cornell University and founder and editor-in-chief of the *Journal of Physical Chemistry*. In the space of a few years around 1930 he proposed a colloid theory of anaesthesia, which escalated into theories of poisoning, drug addiction, and insanity: he believed that all nervous afflictions were caused by precipitation of protein in the nerve fibre and that they could be cured by injecting solutions of salts that were known to solubilize proteins in the laboratory.

His pronouncements hit the front pages of the newspapers, but raised the wrath of the medical profession. Bancroft's work was soon repudiated by his fellow physical chemists and he was stripped of his editorship of the physical chemistry journal. In 1933 the New York section of the American Chemical Society voted to bestow a medal on Bancroft for this work, which raised a storm and embarrassed the society. Bancroft was asked to accept the award explicitly for unrelated and much sounder work that he had done a decade or two earlier; when he refused, the American Chemical Society withdrew the prize altogether—no award was made at all that year. The entire story is told in great detail by J. W. Servos and some of it also appears in Laidler's history of physical chemistry, but to our knowledge there has been no speculation about the malign psychological aberrations that presumably lay behind it.[32,33] The influence of Bancroft on protein science was negligible, possibly because the authoritative figure of A. A. Noyes was consistently scornful of his work and did not even allow the *Journal of Physical Chemistry* in his departmental library at the California Institute of Technology.

The true prophet, young Wolfgang Ostwald, entered the scene around 1907, and did so by the biological pathway He had studied zoology in his native Germany and then went off to spend two years (1904–1906) in Berkeley, California, with Jacques Loeb, spectacular leader of a new mechanistic school in American biology.[34] Just at that time, Loeb toyed briefly with colloid science as a possible source of mechanisms and expounded on it in his lectures, though he quickly became a critic instead. But Ostwald, once infected by the colloid virus, was immune to the subsequent criticism: he became obsessed with the notion of secondary aggregation as a general mechanism for all colloids and pleaded for it with missionary zeal. He returned to Leipzig fully committed to establishing colloid chemistry as a new branch of physical chemistry. He published a famous manifesto in one of the first issues of the then newly founded journal *Kolloid Zeitschrift*;[35] he then effectively took control of the journal in its second year and served as editor for many years. He not only espoused the notion of colloidal association (with no exceptions allowed—no colloidal particles could be equated with 'molecules'), but went even further to claim that the ordinary laws of physical chemistry don't apply to colloids. His enthusiasm was unbounded and he fully supported the journal's conviction, as expressed in its first issue,[36] that colloid science would prove to be important to virtually all fields of endeavour:

But also the fundamental chemical and physical investigation of colloids is absolutely essential for the further development of the individual sciences themselves, particularly geology, biology, physiology and botany. The entire life of the cells (in the animal and plant realms) and of inorganic nature meet in large part at the interactions between colloids.

Ostwald preached his message far and wide, mostly lecturing, not experimenting. In one tour in the United States in 1913–1914 he gave 56 lectures all over the country in 74 days! In his insistence that colloid chemistry has a 'right to exist as a separate field' and in the promotion of his new journal he was displaying an almost pathological desire to emulate his famous father, who, as noted earlier, was a founder of physical chemistry and had launched the *Zeitschrift für physikalische Chemie*. Ostwald's sister, in a biography of her father,[37] speaks of the son's longing to found a 'scientific family' and provides insight (sometimes humorous) into his personality. 'Wolf did not possess the Apollonian composure of my father,' she wrote, 'rather his was a Dionysian temperament.'*

Reaction of protein chemists: 1900–1920

The historian, Robert Olby,[6] has said of this period: 'By common consent the path to the study of metabolic processes and to the structure of natural products lay through the pastures of the colloid chemist.' In fact, this statement does not apply to the group of scientists who, despite diverse backgrounds, had undertaken the hands-on study of the chemistry and physical properties of proteins. Nor does it apply to writers of textbooks specifically about proteins. We shall see that the influence of colloid science on this group was minimal. The macromolecular concept was already an intrinsic part of their framework, usually assumed without need for justification. We can read what was said on the subject around 1900 (before the paper by Noyes, before the Ostwald onslaught) and find it virtually unmodified ten or twenty years later, after the colloid cult's positions had surely become known to every chemist and physiologist. The colloid/macromolecule controversy was mostly simply ignored.

---

* Wolfgang Ostwald finally conceded the existence of true macromolecules in 1930. Colloid chemistry, of course, continues to flourish, mainly as applied to systems (like detergent micelles or inorganic colloids) where 'peculiar physical aggregation' is a legitimate concept, and also extended to aerosols and other situations remote from the original controversies. The *Kolloid Zeitschrift* changed its name to *Kolloid Zeitschrift und Zeitschrift für Polymere* in 1962.

In 1927 F. G. Hopkins[38] called biochemists 'a race apart'; one could well apply these words to protein chemists during the period 1900–1920.

We have already referred to early efforts in determining molecular size from elemental analysis; the data were becoming increasingly reliable as a result of methodological improvements, but by 1900 it was also recognized that one could go far beyond simple elemental analysis. For example, three forms of nitrogen could be distinguished[39] and could be used to provide a new level of characterization. The famous German physiological chemist Alfred Kossel (1853–1927)[40] presciently proposed that amino acids *per se* and their *spatial arrangement* within the protein[41] must become the chemical key to understanding of proteins. Kossel emphasized the same point even more explicitly in an address given in the United States in 1912.[42] The macromolecular nature of proteins, with ordinary bonding between its parts, is at least implicit in all papers of this sort, although many of them were published before the watershed of the 1902 peptide bond proposals. Colloidal theories are never mentioned.

The American chemist Thomas B. Osborne (1859–1929)[43] was (viewed retrospectively) head and shoulders above most of his contemporaries: compulsive attention to meticulous purification, reproducibility, error analysis, etc., shine through all his work. Although most of his work was carried out with seed proteins (being employed by the Connecticut Agricultural Research Station), his results had far-reaching significance. In an extensive paper (1902) on sulphur analysis[44] he included results for animal as well as vegetable proteins, 24 altogether, all yielding minimal molecular weights of about 15 000. Another paper exploits the new ability to differentiate more than one form of nitrogen and recognizes the diversity of amino acids within a given protein—inevitably leading to high-molecular-weight values if integral stoichiometry is to apply.[45] An earlier paper on acid-binding capacity came to a similar conclusion. Osborne is articulate and quite unambiguous about use of the term 'molecular weight' exactly as we understand it today.

Osborne published a monograph[46] on vegetable proteins in 1909, now considered a classic in the field, from which it is clear that the colloidal theory, which had become prominent in the interim since the foregoing papers, raised no question in his mind regarding the macromolecular nature of proteins. In fact, he goes further than before, emphasizing that most proteins probably have molecular weights many-fold larger than the minimal values obtained from elemental analysis. He also asserts that proteins are composed of amino acids, 'united in the molecule with the elimination of water' and that additional amino acids must surely yet be discovered to top up the total analytical recovery from protein hydrolysis to 100 per cent. Nowhere in this monograph does he mention the word 'colloid', nor does he refer to the bevy of colloid chemists who claimed that proteins were aggregates of small molecules not held together by true chemical bonds. It cannot have been out of ignorance: Osborne had until 1903 published most of his important papers in the *Journal of the American Chemical Society*. The 1904 landmark address by the society's president, A. A. Noyes (see above), was of course published in the same journal. Osborne himself became president of the American Society of Biological Chemists in 1910 and was elected to the US National Academy of Sciences in the same year. There are no indications in his career history to suggest that he cut himself off from outside contacts or influences.

Another famous figure, S. P. L. Sørensen (1868–1939), entered the field in the early 1900s. He was one of a succession of distinguished protein chemists at the Carlsberg Laboratory in Copenhagen[47] and is best known for the invention of the pH scale and studies of its physiological and bio-chemical effects.[48] Unlike Osborne, he was prepared to take direct issue with the colloid chemists in general and Wo. Ostwald in particular. Using ovalbumin, he demonstrated the applicability of thermodynamics to macromolecular substances (Gibbs' Phase Rule was obeyed), contrary to the dicta of colloid enthusiasts.* He developed a new osmometer and with it obtained a molecular weight of 34 000 for this protein.† 'Colloidal

---

*J. Willard Gibbs (1839–1903) was the great genius of so-called 'classical' thermo-dynamics, a construct of impeccable logic and absolutely rigorous mathematics, by means of which the laws of equilibrium (chemical, physical, biological) were first established, on the basis of energy and entropy and their associated 'first' and 'second' laws. The laws were all-inclusive, covering systems with any num-ber of components, in phases that were in free contact (liquid, solid, etc.) or in partial contact (semi-permeable membranes allowing only restricted communi-cation). Protein scientists understood the unarguable credentials of these laws and were guided by them; colloid scientists thought themselves somehow exempt.

† The quantitative explanation of osmotic pressure was one of the triumphs of thermodynamic theory. Sørensen's measurement was repeated by Güntelberg and Linderstrøm-Lang thirty years later. They obtained 45 000 for the molecular weight, essentially the value now known to be correct. Sørensen's smaller value was shown to arise mostly from uncertain extrapolation of his data to high dilu-tion and not to error in the measurements *per se*.[49]

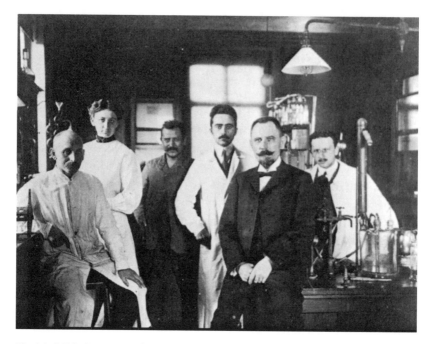

Fig. 4.1 S. P. L. Sørensen and his research group in 1909. (Source: Carlsberg A/S, Archives.)

chemistry has, in my opinion, not contributed to further progress, but rather the reverse,' he said in 1915, in the first of a series of papers reporting on this work.[50]

A more sophisticated paper[51] based on thermodynamics merits explicit mention. It was a youthful effort from Cambridge physiologist A. V. Hill (subsequently a sage of muscle energetics), working in collaboration with Joseph Barcroft, famous for his work on respiration. They measured the effect of temperature on the equilibrium constant for binding of oxygen to haemoglobin, from which classical thermodynamic equations gave them the heat evolved when one mole or molecule of O_2 binds to one mole or molecule of haemoglobin. They also measured the heat directly, per gram of protein, using a calorimeter. The ratio of the two results gave them the molecular weight of the protein, which proved to be close to the familiar figure of 16 000, a minimal value since the stoichiometry of only one O_2 bound had been assumed. There is no question in their mind that this is the true molecular weight—no mention of colloid chemistry or their colloid-oriented Cambridge colleague William Hardy.[52]

One can find isolated advocates of the contrary point of view—colloidal aggregation rather than normally bonded macromolecules—among scientists close to the protein field, but the only significant example was Emil Abderhalden, not actually a protein chemist but an organic chemist studying peptides, who had been a postdoctoral associate with Emil Fischer around 1906. He argued against the importance of Fischer's peptide bond, as we have already mentioned, but this was only part of an overall commitment to the colloid theory. Abderhalden became a founding member of Ostwald's Colloid Society in 1922 and increasingly obstreperous thereafter.[53–55]

Astonishingly, Emil Fischer himself, who had earlier done so much to solidify the macromolecular concept, wavered near the end of his career, venturing the opinion in 1913 that proteins would prove to have a maximal size of about 4000.[56] He presented no rationale for such an extraordinary statement, nor had he previously indicated that he thought there was any limit to the length to which a polypeptide chain could grow.[57]

Turning now to textbook writers in the same period, we note that there were many of them, attesting presumably to an escalating general interest in proteins. The list includes Otto Cohnheim (1873–1953), who was to go on to a long and successful career as physiologist and enzymologist, but who wrote a popular book on protein chemistry in 1900,[58] while still in his 20s. Gustav Mann published an English version of the book in 1906, which includes some of Mann's own views where they differed from Cohnheim's. Colloid chemistry is mentioned here, but without suggesting any importance in regard to molecular weight: for example, to quote directly, 'there can be no doubt that the colloidal albumins possess an extraordinarily high molecular weight'. Two short books about proteins,[59] published in England in 1908–1909 and intended to be introductory texts, pay virtually no attention to colloidology at all. The most passionate on the subject was T. Brailsford Robertson (1884–1930), professor of biochemistry at the University of California, dedicated especially to the application of physical chemistry to his discipline. He wrote an important textbook, which acknowledges that proteins are typical colloids by Graham's original definition, but then goes on to say that the only common characteristic of the group is an 'enormous' molecular weight and volume and that nothing is gained by trying to force them all into 'artificial conformity'.[60]

Emil Abderhalden, already mentioned as a colloid enthusiast, was of course on the other side. He was the author of an influential textbook on the broad topic of physiological chemistry, but proteins form a large part of the subject matter.[61] In this book, even the high molecular weights of haemoglobin calculated from the iron content are discounted. We should also mention Wolfgang Pauli (1869–1955), father of the more famous physicist of the same name (discoverer of the 'exclusion principle'), who more or less switched sides between 1906 and 1912. In a book written in 1902,[64] he was dedicated to the colloidal association theory, but later, in a book published in 1922, said to be based on lectures delivered in 1912–1913, he had changed his mind. Colloid chemistry has evolved, he says; many of its original dicta must be seen 'in a new light'.

The conversion of T. Svedberg

Ironically, the nail in the coffin of the colloid aggregate theory of protein structure came from a scientist within their own ranks. Theodor (commonly known as 'The') Svedberg (1884–1971) began his professional life as a colloid enthusiast, with a doctoral dissertation on that subject.*

In 1909 he wrote a monograph on the preparation of inorganic colloids and in 1921 he wrote a follow-up in English, all within the conventional framework of the colloid theory.[63] Proteins are mentioned only once in the 1921 book, among other colloids found in plants and animals; they are described as 'highly disperse', formed under as yet unknown conditions. Svedberg designed and developed the ultracentrifuge explicitly to investigate the heterogeneity of all colloidal solutions—his first projects were to study inorganic colloids, especially gold sols.

A recollection contributed by John Edsall (personal communication) illustrates Svedberg's opinions at the time and also reinforces what we have already demonstrated, that protein chemists had confidently espoused a contrary macromolecular view long before Svedberg arrived on the scene. The story relates to a lecture given by Svedberg several years

---

* Data in Svedberg's dissertation on diffusion of particles of colloidal platinum prompted Albert Einstein to publish a classic note on the theory of Brownian motion. Svedberg was clearly from the very beginning a man to whom attention was paid. See Einstein.[64,65]

Fig. 4.2 Theodor Svedberg. (From Svedberg's obituary in *Biographical Memoirs of Fellows of the Royal Society*. By permission of the President and Council of the Royal Society.)

later in the laboratory directed by Edwin Cohn and John Edsall at Harvard University.

> Svedberg began with some personal recollections of a visit Edwin Cohn had paid him in Uppsala about 1921 when Svedberg was in the long process of developing his ultracentrifuge so that it would really work reliably. Cohn had been working for a year or more in Sørensen's lab in Copenhagen. He came to see Svedberg before he returned home. Svedberg said he asked Cohn at that time: 'What sort of picture do you think I will get when I have the ultracentrifuge working properly, and put a purified protein solution in it?' and Cohn replied 'I think you will see a single moving boundary in the centrifugal field.' Svedberg said to the group at the seminar that he had, with his background in colloid chemistry, disagreed. He thought that there would be multiple boundaries from different components of such a colloidal substance. He went on to say 'Of course, when I started to do experiments on a purified protein [haemoglobin] I saw at once that Cohn was right and I was wrong'.[66]

The first protein studied by Svedberg (in 1924) was casein from milk—
now known to be a mixture of different proteins—and it produced no
surprises for him. But then came haemoglobin[67] and with it the shock.
Haemoglobin was not only monodisperse but sedimented with a molecular weight of 67 000, four times the by then fully accepted minimal
molecular weight based on elemental analysis with one Fe atom per
molecule.*

The inference, of course, was that the protein in solution is formed by
secondary association of 16 700 subunits with one Fe atom apiece and
even here colloid theory is confounded. Svedberg and Fåhraeus, in their
own words, found 'lack of any systematic variation of molecular weight
with distance from the centre of rotation'—that is, as a function of
increasing concentration. Laws of equilibrium dictate that degree of association should increase with concentration in any reversibly associating
system, and the result thus runs counter to the expectation of heterogeneity that was supposed to typify colloidal systems. (Svedberg went on
to extend his work over the ensuing years to numerous other proteins,
with similar results for the vast majority of them. His ultracentrifuge
became a standard fixture in every protein chemist's laboratory.[68])

In evaluating Svedberg's work, it must be appreciated that the design
and construction of the ultracentrifuge had represented a huge techno-
logical challenge. High speed of rotation was only the beginning. The
speed had to remain constant for many hours, constant temperature was
required as well, and so was high stability—not the slightest wobble of
the rotor could be tolerated because that would stir up the sedimenting
solution. And the *progress of sedimentation* had to be followed constantly
in situ while the machine remained running. One doesn't know which to
admire more: Svedberg's technical skill or the mental ability to discard
well-cherished traditions almost instantly when experimental data
demanded it. When Svedberg received the Chemistry Nobel Prize in 1926,
the decision to make the award seems to have been based on the instru-

---

* G. S. Adair had measured the molecular weight of haemoglobin by osmotic
pressure at about the same time and got the same result. Svedberg did not know
this, of course, being at that point not part of the protein 'network'. Adair's result
undoubtedly enhanced credibility of Svedberg's result, obtained by what was
then an utterly novel technique. See Chapter 9 for more detail about Adair's
work.

mentation alone, before the significance of his haemoglobin work was fully grasped. The presentational speech from the Royal Swedish Academy[69] still waxes enthusiastic about the virtues of 'colloid chemistry' as an intellectual means to think about naturally occurring substances. Svedberg's own talk at the award meeting was full of illustrations that provide details of apparatus design.

Staudinger and the confrontation

Scientists can often occupy different worlds depending upon their disciplinary bent. Polymer chemists, unlike protein chemists, had no preconceived faith in the existence of huge molecules, with monomer units chemically bonded into very long chains. Proposals for the structure of both synthetic and natural polymers did exist that involved some sort of association of small units by means of secondary interactions (non-valence bonds). In the case of rubber, cellulose, and synthetic polymers there was no reason not to take them seriously. The celebrated conflict between Staudinger and the 'aggregationists', described in detail in the references cited at the beginning of this chapter,[3-8] did indeed take place.

A meeting of the Versammlung der Deutschen Naturforscher und Ärtze in Düsseldorf in 1926 is memorable in the lives of the polymer chemists who participated as the occasion of the key confrontation of Staudinger's campaign, where some of his reluctant colleagues became convinced. It is sometimes claimed that biochemists were present and arrayed against Staudinger's position, but the record[70] shows only two biochemical participants: Max Bergmann, who qualifies as a protein chemist, and Ernst Waldschmidt-Leitz, an enzymologist.[71] Both were eccentrics, for most of their careers outside the mainstream. Waldschmidt-Leitz, for example, claimed as late as 1933 that enzyme activity resides not in specific proteins, but in low-molecular-weight compounds 'adsorbed' to non-specific protein carriers.[72]

The most revealing evidence for the existence of two different worlds comes from the lack of cross-references. Staudinger, in his talk at the Düsseldorf meeting,[73] makes no mention of Svedberg, who received his Nobel Prize in the very same year of 1926. Svedberg in turn, in his Nobel award address,[74] which of course focuses on proteins, mentions other organic polymers of high molecular weight (rubber, starch, cellulose), but does so without any reference to Staudinger. Staudinger himself seems never to have became aware of the protein picture. When he received his own Nobel Prize in 1953 for 'discoveries in the field of macromolecular chemistry', he acknowledges[75] Svedberg's 'astounding discovery' in 1926 that some proteins are naturally occurring macromolecules of high molecular weight, but goes on to say that positive proof 'has so far been found in only a few cases'—which for 1953 is surely ludicrous.

CHAPTER 5

Bristling with charges

Of the proteins that have been studied thus far, excepting only the albumins, oxyhaemoglobin has the steepest slope to its titration curve in the neighborhood of its isoelectric point. Its titration curve, from pH 6 to 7.5, is most readily interpreted by assuming the presence of both positive and negative charges upon the oxyhaemoglobin molecules throughout this range.

E. J. Cohn and A. M. Prentiss, 1927[1]

Introduction

Before Arrhenius it would have been unthinkable that there might be bare ionic charges, let alone positive and negative charges side by side, sprouting out all over a polypeptide chain. Electrostatic attraction had been the linchpin of all theories of chemical affinity, the only sensible idea available; opposite charges were seen as inexorably seeking each other out, forming chemical bonds in the process. The ultimate goal was not precisely an electrically neutral world, for positive and negative charges with a range of values were thought to exist and would not necessarily balance out in compound formation, but there was no rival to the electrical force itself. Humphry Davy had believed it, so had Berzelius. This viewpoint remained strong enough in 1860, long after both Davy and Berzelius were dead, to produce an impasse at the famous Karlsruhe conference to define a table of atomic weights. To arrive at self-consistent values would require formulas of H_2, O_2, etc., for some of the most familiar gas molecules, and that was unacceptable to many—identical atoms necessarily carried identical charges and lacked any conceivable mechanism for combination.[2]

Svante Arrhenius, of course, put an end to this in 1887 with his theory of the electrolytic dissociation of salts in water into their separate ions, which was almost at once quite generally accepted—Wilhelm Ostwald's *Lehrbuch der Allgemeinen Chemie* (published in the same year, 1887) includes the theory as textbook material.[3] But the intuitive 'certainty' that some sort of attraction between opposite charges must remain persisted for a long time. Arrhenius's theory was thought to apply fully only at great dilution and recombination of cations and anions was thought to take place at higher concentration, as ions came on average closer to each other. And on a single protein molecule, where chemical bonds would ensure close proximity regardless of concentration and where, furthermore, water cannot by any means fill all the intervening space, uncombined charges of opposite sign were surely at best wild speculation—at least, that was the dominant opinion.

Physiologists and others with biochemical interests quickly adopted the Arrhenius theory in principle; for example, Jacques Loeb, at a meeting held in December 1897 to outline the biological problems of the day, said:

> The theory of the dissociation of electrolytes is of fundamental importance in the analysis of the constitution of living matter . . . . The specific effects of inorganic acids are due to the number of positively charged hydrogen ions in the unit of solution, and the specific effects of alkalies to negatively charged hydroxyl ions.[4]

Loeb (1859–1924), German-born, but an early immigrant to America, was a physiologist at the time, a commanding figure in his field. He would soon follow his own precepts and turn to physical chemistry as his favourite realm. His 1922 book, *Proteins and the theory of colloidal behavior*, already mentioned earlier for its opposition to theories of colloidal association, also provides a clear and exciting discussion of the issues and controversies that became involved in the electrochemistry of proteins.[5] Loeb himself may not have seen the final outcome. He died in 1924, a year or two before the firm establishment of the ultimate model of the multi-ionic state of proteins.

Zwitterions: Bredig to Bjerrum, 1894–1923

We begin with a purely chemical extension of the ionic theory *per se*—the notion of a dipolar molecule, with no total charge (no contribution to electrical conduction) but containing within it a full positive and a full negative charge. Such a dipolar molecule is often named *zwitterion** and was first envisaged in 1894 by the chemist Georg Bredig, a student of Wilhelm Ostwald's at the Physical Chemistry Institute in Leipzig. Bredig, in a study of the dissociation of a huge number of organic bases, concluded on the basis of simple chemical reasoning that one of them, the trimethyl-ammonium derivative betaine, must in its neutral form be what he called an 'internal salt', with the formula $(CH_3)_3N^+-CH_2-COO^-$.[6] Simple amino acids in their state of net neutrality were soon proposed to be dipolar ions, too—that is, glycine could be $NH_3^+-CH_2-COO^-$ and not NH_2-CH_2-COOH. [7]

Although both Bredig and his Leipzig colleague K. Winkelblech were explicit in saying that a plus and a minus charge coexist on the same molecule, there was a general reluctance to admit that these ionic charges extend freely from the amphoteric molecule into the adjacent solution. The term 'inner salt' seems to have been visualized by most people as implying that the two opposite charges were actually in neutralizing contact to form closed intramolecular rings. There was much confusion on the subject, fuelled by the prevailing custom of treating acids and bases as *formally* distinct, which often led to the incorrect assignment of observed acid–base reactions.[10] This and some other questions of mathematical formalism were completely clarified in 1916 by a California chemist, E. Q. Adams, in the course of an insightful general treatment of all kinds of acids and bases with multiple functional groups.[11] Adams quite rightly castigated the formal distinction between acids and bases (*dissociation* of OH^- is in fact equivalent to *binding* of H^+) and he was the first to use the now familiar thermodynamic 'square' for analysing molecules with more than one acidic group. Among other applications, he explicitly diagrammed the 'square' for glycine as given in Fig. 5.1.

---

* The actual word 'zwitterion' was coined by the German chemist, F. W. Küster, based on the German word 'Zwitter', meaning 'hybrid', and it was eventually absorbed into English as well. Küster[8] was studying acid–base indicators and had no interest in amino acids. He concluded that the acid form of methyl orange must be the dipolar $NH(CH_3)_2^+-C_6H_4-N_2-C_6H_4-SO_3^-$ on the basis of colour changes accompanying the acid–base transitions, comparing them with data for appropriate model compounds. This is one of the earliest examples of the application of spectroscopy to a purely chemical problem. It would be nearly 40 years before a similar application of spectroscopy became possible for glycine and other amino acids, where infrared and Raman spectroscopy had to be used because of lack of visible 'colour'. See J. T. Edsall[9].

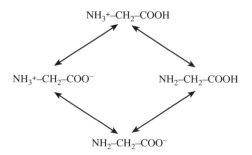

Fig. 5.1 Diagram of the two conceivable pathways for the acid dissociation of glycine. Given the likely values for the dissociation constants of the constituent groups, the predominant neutral form unquestionably had to be the zwitterion: specifically, $[NH_3{}^+\text{–}CH_2\text{–}COO^-]/[NH_2\text{–}CH_2\text{–}COOH]$ had to be greater than 10^6.

But Adams, though a rather brilliant chemist, was far-ranging in his interests, publishing papers on a variety of topics, and never became well-known. In particular, his model for glycine apparently failed to catch the attention of anyone primarily interested in proteins or amino acids, which may have been because the work was published in the midst of World War I. It was after the war, in 1923, that the Danish electrochemist Niels Bjerrum reiterated the same principles, this time with explicit focus on amino acids and peptides. Only then did the zwitterion notion begin to have a significant effect on the segment of the scientific population on which this book is focused.[*][12]

Motion in an electric field

We have focused in the previous chapter on the gulf between colloid chemistry and much of the rest of the scientific world. To many the distinction between 'colloids' and 'crystalloids' pointed the way to the promised land; to others it must have seemed like madness. However,

---

[*] One should perhaps remind the reader that glycine itself is not a valid model for any part of a protein molecule, because it exists there in peptide linkage. Proteins are amphoteric because they contain amino acids with acidic or basic side chains, which (at appropriate pH) retain their charges in peptide linkage. Among the amino acids listed in Table 3.1 of Chapter 3, arginine, histidine, and lysine potentially carry positive charges, and aspartic acid, glutamic acid, and tyrosine potentially carry negative charges.

everyone was agreed on one point, namely that colloidal particles are electrically charged. Colloidal gold, for example, was actually prepared by an electrolytic process, the soluble particles, obviously charged, being sloughed off from gold electrodes; other inorganic colloids were aggregates of normally insoluble salts and ionic charges were again an essential element in understanding how stable soluble particles could be formed.

Proteins were 'colloids', sometimes included in broad experimental projects, and the fact that proteins exist as charged particles in solution came directly from the pioneering work of two British colloid chemists, H. Picton and S. E. Linder, who, in 1892, were the first to demonstrate electric charge directly by measuring the movement of colloidal particles under the influence of an electric field. They mainly investigated inorganic substances like colloidal arsenious sulphide, but also included the crystalline protein haemoglobin and remarked on the ease with which one could follow the movement of haemoglobin molecules in solution, as represented by the moving boundary between the red protein solution and colourless solvent. In effect, this was the origin of moving boundary electrophoresis and the precursor of methods such as gel electrophoresis, which are so much a part of today's protein chemist's toolbox.[13,14]

Picton and Linder's work was extended in 1899 by the Cambridge colloid chemist William Hardy, who showed that protein particles could move in opposite directions in an electric field, depending on acidity, 'with the negative stream if the reaction of the fluid is alkaline; with the positive stream if the reaction is acid'. In other words, a protein particle could be positively charged at low pH, but negatively charged at high pH, and switch reproducibly from one form to the other.[15]

This was a momentous finding by any criterion, only a decade after free ions of any kind were first postulated, for all ions, including charged colloidal particles, had naturally been assumed to be either cations or anions; a transition from one to the other opened up entirely new fields of investigation. It is somewhat puzzling today to find that the protein used by Hardy was derived from a *boiled* solution of egg-white—hardly a good example of a native protein—but similar results were soon obtained for more gently treated material. Pauli, for example, showed results just like Hardy's for undenatured serum albumin.[16]

One consequence of variable charge was the necessary existence of a point at which the motion changed direction, which was called the 'isoelectric point'—the term still used today. Its measurement was a matter of particular interest, both theoretically—how should this point be viewed at the molecular level?—and also practically, for the isoelectric point was usually a point of minimal solubility for each protein and thus a vital factor in the quest for improved purification procedures. Leonor Michaelis, subsequently famous for his numerous contributions to the interface between biochemistry and physical chemistry, made a particularly accurate measurement for serum albumin: at an acidity of 2.1×10^{-5}, the protein behaved as a cation, at $[H^+] = 1.9 \times 10^{-5}$ it behaved as an anion. The pH scale was not yet in use; the corresponding pH values would be 4.68 and 4.72, respectively—that is, the isoelectric pH is 4.70 ± 0.01.[17] This is a good illustration of the great accuracy of many measurements that were beginning to be made on protein solutions at the time (1909).[18]

Titration of proteins

Proteins are amphoteric, able to function as both acids and bases, a property obviously related to the effect of pH on their mobility in an electric field. The phenomenon can be investigated by 'titration'; that is, by the progressive addition of measured amounts of acid or base and the determination of how much is bound to the protein and how much remains free—a quantitative accounting of how many actual charges there are as a function of pH, much more informative than the mere measurement of the direction and rate of motion in an electric field.

Detailed investigation had begun already before Hardy's electrophoretic experiment, with a classic paper published in 1898 by two Hungarians, Stefan Bugarszky and Leo Liebermann, who titrated purified egg albumin.[20] They were the first to study *both* acid and base binding, and the first to apply the state-of-the-art physical technique of using electromotive force to determine hydrogen ion concentration, a method just recently devised by Walther Nernst[21] and far more elegant than other methods then in common use.* They also applied proper thermodynamic analysis to their binding data, demonstrating that binding was a true reversible equilibrium process, leading to saturation as the extent of binding increased.

---

* The enzymologist Otto Cohnheim had used the rate of sugar inversion as an indirect measure of H^+ concentration; Thomas Osborne, following Bugarszky and Liebermann's work in 1902 with a titration study of edestin, still relied on the crude use of colour-changing indicators.[22]

Bugarszky and Liebermann, of course, did not try to identify the sites responsible for the binding (we must remember this was only 1898, and even the existence of the peptide bond had not yet been proposed) but their work was soon taken up by others, who had no doubt that the sites must be derived from a protein's constituent amino acids, chiefly the acidic carboxyl groups located on aspartic and glutamic acid residues and the basic amino groups from lysine and histidine.

This was, for example, the completely accepted view at the Carlsberg Laboratory in Copenhagen in 1917, when Søren Sørensen published his series of studies on proteins, to which we made reference in the previous chapter.[23] Kai Linderstrøm-Lang, a former pupil of Sørensen and later (in 1938) his successor as director of the Chemistry Department, expressed it as follows (in a famous paper to which we shall shortly return):

> The fact is simply this, that in egg albumin . . . we have a substance with many acid and basic groups, whose dissociation constants do not lie far apart, and whose dissociation curves therefore partly overlap and efface each other. This idea is not new, and has often been impressed upon me by Sørensen himself.[24]

Incredible precision was attainable in these measurements: protons were literally being counted one by one as they combined with a protein or dissociated from it. Famous figures like Jacques Loeb and Leonor Michaelis, as well as less prominent ones, became involved and measurements were made for a variety of proteins. There was broad agreement between different investigators and the number of sites required to account for the results was appropriately large; for example, maximum binding was around 50 H^+ or OH^- per molecule in the case of egg albumin. An extensive review and comparison of data from numerous laboratories was provided in a 1925 review by Edwin Cohn.[25]

However, there was a snag, an unanswered question. What is the correlation between titration range and amino acid identity? The underlying problem was the same one that we have pointed to before in discussing simple zwitterions, namely the obstinate persistence in treating acids as distinct from bases. Just as this led to failure to recognize the zwitterionic form as the dominant species of the neutral glycine molecule, so it also led to misunderstanding of what was going on in proteins.

Jacques Loeb, the great promoter of the importance of understanding ionic interactions, was himself a typical example. Proteins were unquestionably viewed as amphoteric, capable of binding both acids and bases, but in the isoelectric state they were *assumed to be uncharged* and this was interpreted as meaning that individual $-COOH$ and $-NH_2$ groups were all uncharged.[26] When acids such as HCl were added to the isoelectric protein (lowering the pH), the acids were thought to bind to basic sites, forming salts of the type $-NH_3{}^+Cl^-$. Binding of base on the alkaline side was likewise interpreted as formation of COO^-Na^+. Adams had in principle demonstrated the fallacy of all this in 1916, but protein chemists (like ordinary organic chemists) were either unaware of this or unconvinced—not till Bjerrum's 1923 paper did the light dawn.

In this case we can actually pinpoint almost the exact date when one of the leaders in the field, Leonor Michaelis, became convinced. Michaelis was the author of a popular German textbook on H^+ and its reactions, the second edition of which was translated into English in 1926.[27,28] Direct translations do not normally involve substantive change, but in this case Michaelis felt compelled to make a few alterations to reflect significant recent advances. The 'most marked advance in the theory of dissociation', according to Michaelis, has been the clarification of the concept of zwitterions by Adams and Bjerrum. He points out that 'an ion is now the undissociated species'—$^+NH_3-CH_2-COO^-$ predominates over NH_2-CH_2-COOH. He has clearly just discovered this for the first time. All available evidence indicates that the timing of his personal enlightenment applied to protein science in general. (With the exception, presumably, of people like Abderhalden[29], who still wandered in the wilderness of speculative organic chemistry and denied the validity of the peptide bond.)*

---

* The German physical chemist Hans Hermann Weber (1896–1974), best known for his work on muscle proteins, gave the first convincing *direct* demonstration that isoelectric proteins are zwitterions—truly bristling with charges (+ and – in equal number). By 'direct' we mean that it was based on experiments with proteins themselves and not on analogy with small molecules. Weber made calorimetric and volume measurements as a function of pH, as proteins were titrated from one side of the isoelectric point to the other. He knew that the heats of dissociation of H^+ from amino groups are generally much larger than those for COOH groups and that the converse is true for volume changes accompanying dissociation. His results unambiguously identified the titrating groups on the acid side as being mainly COOH and those titrating on the alkaline side as being mainly amino groups. In the isoelectric state, the former would then be mainly COO^- and the latter $NH_3{}^+$.[30]

Spherical shape

Thus proteins came to be seen, at least by everyone with some appreciation for a physico-chemical approach, as multi-charged ions, able to
accommodate very high charges per molecule. At first the ions were
probably thought to be all of one sign, positive on one side of the isoelectric
point and negative on the other, but increasingly, as the zwitterion principle sank in, it was realized that both positive and negative charges could
be accommodated simultaneously.

At this point new questions arose. One needs to know something about
a protein's general shape. How are the charges distributed? Problems of
this kind had not been previously addressed with specific reference to
polypeptide chains. When Emil Fischer synthesized his decamers and
more, it was not done in an atmosphere that was ripe for speculation on
how the chains might disport themselves in solution.

In practice, those who needed to have an answer took their cues about
physical concepts like particle geometry from the realm of colloid chemistry and the simplest guess was that protein molecules in solution would
resemble colloidal particles in this respect. Colloidal gold could not conceivably be anything other than densely packed spherical particles, with
an electrically charged surface ('adsorbed' ions) to keep them in suspension or solution. It was plausible that protein molecules would likewise
be spheres, with multiple charges bound to the surface to give them solubility in water. It was known that some proteins could swell by imbibing
water and the mental picture of protein molecules would therefore have
differed from the picture of colloidal gold by allowing the particles to be a
little larger than one would infer from their molecular weights alone. But
there is no evidence that anybody thought proteins might be flexible
chains, meandering all over the solution, a concept that would later
dominate thinking about synthetic organic polymers.

One who particularly needed to make an assumption about molecular
shape was Kai Linderstrøm-Lang. Linderstrøm-Lang (1896–1959) was at
the start of his career, impressed by the power of physical chemistry as an
investigative tool, eager to relate the protein ions to the mainstream of
the physico-chemical theory of electrolyte solutions.[31] The latest advance
on that subject was the Debye–Hückel theory for the statistical distribution of free ions about a central ion, which had just been published in
1923, and Linderstrøm-Lang brilliantly adapted it within a year (remark-

Fig. 5.2 Kai Linderstrøm-Lang. The stubby object in Lang's right hand is a cigar—he was rarely seen without one. (Source: Carlsberg A/S, Archives.)

ably speedy assimilation) to a multi-charged central ion, such as a protein molecule would be.[24] In so doing, he had to visualize the protein molecule as a body that could be represented in mathematical terms, and he did what probably anyone else at the time would have done, he unhesitatingly chose a sphere. He gave the sphere a density about equal to that of a solid protein, and on this basis calculated its radius from the protein molecular weight. He felt no need for justification. Linderstrøm-Lang obtained 22 Å as the radius for egg albumin in this way,[32] an astonishingly small value, though still, of course, considerably larger than the radius of water molecules or ions such as Na^+.

Many years earlier (probably not known to Linderstrøm-Lang) there had actually been an experimentally derived estimate of similar compact size. It came from an adventurous paper, not often referred to, by an Australian theoretical physicist, William Sutherland (1859–1911), and it was communicated to the Australian Association for the Advancement of Science in 1904 and published in England a year later. Sutherland (unusually for a physicist) was familiar with the difficulty in measuring protein molecular weights and knew that most of the available data were only minimal values. He suggested that diffusion could provide a solution to the problem, by use of Stokes' law for spheres, which relates the diffusion coefficient directly to the radius of the sphere. He went back to Thomas Graham (!!) to get a diffusion coefficient for egg albumin, which gave him a molecular radius of 19 Å. An estimate of particle density then gave a molecular weight of 33 000, which, of course, is almost exactly the same as the accepted molecular weight in the time of Linderstrøm-Lang.[33]

What is of special interest here is not really how close to the true numbers he managed to get, but rather the attention that Sutherland paid to the errors that might be involved. He was compulsive about considering sources of error that we might today regard as only remotely likely, such as possible slippage between the particle surface and the solvent, that would affect which of two limiting forms of Stokes' law to employ. But the one assumption he did not question (did not even explicitly mention) was the assumption of a solid spherical shape for the diffusing particle! The most plausible conclusion is that he thought that anybody else would have made the same choice.

Direct measurement by viscosity could have been decisive on the question of shape because viscosities were easy to measure (proportional to flow times through a capillary) and data for proteins had been reported as early as 1902.[34] More extensive data came from the laboratory of Harriette Chick at the Lister Institute in London about ten years later.[35] Moreover, Albert Einstein's viscosity equation was available to interpret the results. This equation was specifically derived for the effect of large spherical particles on viscosity. It showed that the viscosity increment should depend only on the volume fraction of the dissolved spheres, with no dependence on the size of individual particles.[36] In the case of proteins, this meant that uncertainties about molecular weight could be ignored; one did not have to know the molecular weight to make use of the equation.

As it turned out, viscosity data had little influence. Einstein's equation was indeed approximately obeyed by crystalline proteins, such as egg albumin and serum albumin, but much larger viscosities (viscosity increments as much as 100-fold larger under some conditions) were found with some other proteins—for example, casein and gelatin. The strong expectation of a compact spherical shape for all proteins may have come into play here. Rather than celebration at unequivocal confirmation for some proteins, there was concern when the opposite result was obtained. Einstein's equation was strictly limited to spheres and could not interpret the contrary result—hydration alone could not conceivably account for the largest measured discrepancies. Arrhenius, in 1917, reviewed the available data and was generally unenthusiastic, more intent on promoting an earlier empirical equation of his own than on using the results constructively for an insight into protein structure.[37]

Equations for the viscosity increment of non-spherical solute particles eventually became available in 1940, but long before then Svedberg's ultracentrifuge convincingly confirmed the assignment of a compact, close to spherical shape on the basis of Stokes' law for frictional resistance. This law is also limited to spheres, but it had been a bastion of fluid dynamic theory since 1847, a familiar tool with which people were more comfortable than with Einstein's viscosity equation. Svedberg's first results from the dual use of the ultracentrifuge (determining both mass and shape) were reported in 1929. Two crystalline proteins, Bence-Jones protein and the perennial egg albumin, having entirely different chemical compositions and different isoelectric points, were found to have virtually identical molecular weights and shape. Molecular frictional coefficients for both were extremely close to what is expected for a perfect sphere. The calculated radii were 21.8 Å and 21.7 Å, respectively.[38]

The globular protein

It is said that a picture is worth a thousand words, but in the 1920s pictures were not yet being employed to supplement the verbal description of structural or mechanistic ideas. Illustrations in chemical books or articles were used for graphical display of experimental results and often to provide drawings or photographs of apparatus, but not to represent speculative visions of molecular structure. Thus, if we ask what people had in mind when they reflected on their own work and that of others, when they tried to create a mental imagine of a protein molecule, we have only the thousands of words to go by and no pictures to verify our conclusion.

On this basis, it is clear that between 1925 and the early 1930s a common overall view became widely accepted as a mental image for many proteins, incorporating both aspects of the model we have been discussing in relation to Linderstrøm-Lang: the compact, close to spherical particle shape and the high density of charges on the surface of that particle. Most of the common soluble proteins—haemoglobin, egg albumin, serum albumin, most enzymes—were seen to embody the main features of this model and became known as 'globular' proteins. The name was based, of course, on molecular shape and charge *per se* did not enter into the rationale behind it.[39] But the term did not come into use until 1934, by which time the multi-polar character of proteins was completely established, and ionization would have been part of the picture for most biochemists. Surface charges and the fact that they were so close to each other went hand-in-hand.

Linderstrøm-Lang himself, in his application of the Debye–Hückel theory, could not specify (even theoretically) the actual coordinates of individual protein charges; he could not even allow for the presence of both positive and negative charges on the same molecule. But this was not a problem unique to proteins and was equally true for dipolar ions such as betaine or glycine. *Net charge*, evenly distributed, was the only parameter the theory could handle and Linderstrøm-Lang admitted that 'we cannot go beyond a rough approximation' in applying the theory to experimental data.[42] But both kinds of charges are implicit in Linderstrøm-Lang's paper by virtue of repeated references to Bjerrum's zwitterion principle—and there can be little doubt that they were part of the popular model.[43]

There is general evidence of lingering uncertainty concerning the number of charges and how close to each other they would have to be, but it did not last long.[43] Edwin J. Cohn, who would come to enshrine the 'bristling with charges' concept for posterity by putting it into the very title of the treatise on proteins he later wrote with John Edsall,[44] still expressed some doubt in a 1925 review[45] but it was gone a year later. In a paper (submitted in July 1926) that is a perfect illustration of the natural alliance between shape and ionization in thinking about soluble proteins, Cohn was unambiguous about the large number of ionic groups and that they are 'highly dissociated both as acids and as bases' at the isoelectric point. The most ready interpretation of the titration curve of haemoglobin in the isoelectric region was to assume 'the presence of both positive and negative charges throughout this range'.[46] A 1932 review emphatically reiterated these principles.[47]

Even the mathematical problem of incorporating actual coordinates of individual protein charges into the Debye–Hückel theory was solved in 1934. It was done by J. G. Kirkwood, one of the foremost theoretical chemists of his day, without vested interest in any particular view of protein structure, and his theoretical extension amounted to an affirmation of the model as we have described it.[48] The theory was (as is typical for any good theoretical work) formulated to be as general as possible. However, it was initially applied only to small molecules, including the aliphatic amino acids, for the purpose of explaining shifts in acid dissociation constants arising from interactions between the constituent charges. Applications to protein titration curves were not made until more than twenty years later.[49]

Fig. 5.3 The Cohn and Edsall research group at Harvard University in 1944. This laboratory was the embodiment of the 'bristling with charges' principle. Note the contrast to the Sørensen group (Fig. 4.1) and the military uniforms testifying to liaison with and financial support from the military. (From the authors' personal papers; originally courtesy of J.T. Edsall.)

Also in the 1930s, dielectric constants of protein solutions were being measured and electric dipole moments of protein molecules were calculated from the data. Reviewing the results in 1938, Edwin Cohn remarked about how small the dipole moments were, given the huge number of charges known to be present.[50] It indicated that the charged groups must be very evenly distributed about the molecular surface. The dipole moment of haemoglobin was shown to be equivalent to what would be generated by just two positive charges at one end of the molecule and two negative charges at the opposite end, whereas it was certain that there must be over 100 charges in all. (The dipole moment of egg albumin is even smaller!)

Some historical writers today give the impression that little of value about protein structure was available before X-ray crystallography appeared on the scene. That is clearly not true. The concepts incorporated in the mental image of a 'globular' protein (compactness, bristling with charges) were accurate as far as they went and had become a permanent feature of the protein landscape many years earlier. Add other reactive groups to the surface, derived from uncharged amino acid side chains, and we arrive at a model on the basis of which biochemists and others were ready to speculate freely about active sites of enzymes, immunological activity, etc. Indeed, many pivotal and subsequently productive ideas of protein structure/function relationships emerged from such 'low-resolution' pictures of protein organization.

CHAPTER 6

Fibrous proteins

R. O. Herzog and W. Jancke found that crumpled cellulose fibres lead to Debye-Scherrer rings in monochromatic x-rays, indicating a micro-crystalline structure. Further x-ray results indicate that the cellulose crystallites are oriented parallel to the fibre axis.

M. Polanyi, 1921, laying the groundwork for
all future fibre research[1]

It had been known for many years that not all proteins were globular proteins—that is, relatively compact, nearly spherical, charged particles. Some (like casein and gelatin) deviated grossly from this picture when studied in solution. Others, notably keratin, silk fibroin, and collagen, were not even soluble in aqueous media and could be seen in associated form under the microscope as elongated fibres. By chemical criteria, they were indubitably proteins, composed of the same amino acids as their more tractable brethren, though sometimes the proportions were unusual. Like the globular proteins, these fibrous proteins bear ionic charges. Jacinto Steinhardt, an American who had some years earlier been a postdoctoral visitor to the Carlsberg Laboratory, carried out acid–base titrations of suspensions of wool keratin, and the resulting curves were similar to those for globular proteins in solution.[2]

Textile fibres

In the first decades of the twentieth century, physiological interest in these fibrous proteins was almost non-existent. Compared with the oxygen-binding powers of haemoglobin or the activity of enzymes, they seemed unlikely candidates for providing insight into the meaning and mechanisms of 'life'! On the other hand, broad semi-technical understanding of fibres was at a high level, reflecting popular interest in the subject,[3] and there was a commercial motivation for investigating them—wool and silk, in particular, were important to the textile industry, which had its own research arm and funds to invest for purely practical purposes. By 1930, however, the interests of technology and the life sciences would intersect. The fibrous proteins would unexpectedly provide vital clues in the quest for structural understanding of how all proteins work—including haemoglobin and the enzymes.

Aside from that, there was a change in the fibre industry, the history of which is exceptionally well documented.[4–6] Synthetic fibres and plastics were in the offing! Synthetic rubber already existed, polystyrene and nylon were on the way. There were visions of huge profits and suitably large investments were made. In Germany, R. O. Herzog,[7] as head of the Institute for Fibre Research at the Kaiser Wilhelm Institute, surrounded himself with an impressive array of talent and in Ludwigshafen 'I.G. Farbenindustrie' set up a totally commercial laboratory where Hermann Mark and Kurt Meyer forged a profitable and lengthy collaboration. In America, another outstanding pioneer, Wallace Carothers, joined Du Pont in 1928 after a brief venture into academia at Harvard. He was given virtually carte blanche at Du Pont to do research that culminated in the synthesis and marketing of nylon. A spurt in activity in the scientific study of natural fibres occurred along with the interest in synthetic ones —predictably so, for the success of synthetic fibres would ultimately depend on comparison with cotton and wool and silk. New methods and techniques were developed, the most significant of which was X-ray diffraction.

X-ray diffraction

The accidental discovery of X-rays in 1895 by Wilhelm Röntgen at Würzburg is one of the grand spectacular events in the history of science. It was hailed immediately as the herald of a new dawn in physics, not only by professionals, who sought the physical origin of these mysterious emanations, but also by laymen, who were fascinated by the spectacular power of X-rays to 'see' into the innards of living bodies. It took some years to discover that X-rays were electromagnetic waves of short wavelength, and then some more years to realize that X-ray wavelengths were of the same order of magnitude as interatomic distances in molecules. And this in turn gave them the potential to 'see' the inside of *molecules*, a much more profound kind of vision for the physicist than the ability to see the skeleton within the body.

Exploitation of this prophesied insight at the atomic level, however, required *crystallinity*, regular interatomic spacings, that would lead to reinforcement between reflected or scattered X-rays—that is, creation of intensity maxima. The theoretical condition for reinforcement is given by the famous Bragg equation,[8] which lies at the heart of all X-ray crystallography:

$$n\lambda = 2d\sin\theta,$$

relating regular repeat distance within the crystal (d) to the scattering angle (θ) at which diffraction maxima are observed. Diffraction maxima occur when $2d\sin\theta$ is any integral multiple of the wavelength λ of the applied X-rays. Setting $n = 1, 2$, etc., in Bragg's equation leads to what are known as first order, second order, etc., reflections. Structures of simple crystals, like diamond or NaCl, with only a tiny number of relevant distances, could be quickly elucidated from such repeat distances alone. But organic crystals (including some for proteins) gave X-ray diffraction patterns that were indeed sharp and well defined, but hopelessly intricate and not interpretable at this stage of development of the technique.[9]

It soon turned out that diffraction maxima (less sharp and impressive than those from true crystals) could be seen whenever there was any structural regularity at all, and this discovery was seized upon by physicists at the Berlin-Dahlem Fibre Institute to examine 'fibres' of all kinds, metal wires as well as textile fibres.[10] A distinct and novel diffraction pattern was found in bundles of ramie fibres (cellulose, chemically a polymer of a two-sugar unit called a *disaccharide*), the problem of interpreting which was assigned to Michael Polanyi, a brilliant Hungarian theoretical physical chemist, who had recently joined the Institute. He showed that the pattern was indicative of one-dimensional order, crystalline regions within the cellulose fibre that were repeated regularly along the fibre axis.[11,12]

In quantitative terms, things were not so simple, for Polanyi's interpretation led to the controversial and highly publicized 'unit cell paradox'. The repeat distances deduced from his analysis were very small and the crystallographic unit cell dimensions based on them were correspondingly small—$7.9 \times 8.4 \times 10.2$ Å for cellulose.[13] A unit cell is the part of a crystal that contains the fundamental repeating unit, normally one whole molecule, or sometimes two or three, depending on symmetry, but Polanyi's dimensions obviously could not conceivably accommodate anything even approaching the size of a macromolecule. Polanyi then suggested that his result was actually compatible with two alternatives: the unit cell could contain, instead of independent molecules, two disaccharide *segments of a long polysaccharide chain*; the more conventional interpretation, based on independent molecules, would in this case have to mean an intrinsic unit of two small cyclic disaccharide molecules. (Many years later, Polanyi lamented that he had lacked the chemical sense to unequivocally support the long chain model!)

A similar result was obtained for silk fibres, studied by R. Brill, a Ph.D. student of Polanyi's, but he limited himself to the conventional explanation that a small crystalline molecule must be responsible for the crystallinity of silk fibroin, the protein constituent of silk. He does not mention his mentor's (Polanyi's) alternative of a single long chain molecule, with a unit cell based on its repeating segments.[14]

Since there was still resistance in some quarters to high molecular weights for proteins (Chapter 4), the small dimensions of unit cells were eagerly seized upon by the colloid side in the colloid/macromolecule debate, as evidence for the notion that large particles always arose from secondary association of small molecules into larger entities. Polanyi's alternative structure was strongly supported in 1926 by two botanists from the University of California (again studying cellulose),[15] mostly on the basis of chemical data, but final acceptance throughout the scientific community did not materialize until after 1928, when Meyer and Mark in the I.G. Farben laboratory proclaimed themselves convinced.[16] X-ray repeats were reinterpreted as 'identity periods' along a fibre instead of being distances between entities unconnected by covalent bonds The unit cell was renamed the 'pseudo' unit of crystallization.

Protein fibres

In the case of proteins, the arrival of X-ray diffraction signalled a profound change in objectives. On its way out was the concept of 'particle shape' (arrangements within the particle unspecified), to be replaced by numerical interatomic distances, the 'd' parameter of Bragg's equation.*

---

* Overall particle shape, with arrangements within the particle unspecified, continued to be important for proteins in solution, as witnessed by extensions of Stokes' law and of viscosity equations from spheres to ellipsoids. 'Axial ratio' as a

Admittedly it was nothing but guesswork at first to identify which particular atoms corresponded to a given distance and it would take a generation before this revolution would take full effect, to produce three-dimensional molecular models. But there is no doubt that it started with fibrous proteins, with the first appearance in print of a polypeptide chain formula that had an interatomic distance marked on it to a precision of close to 0.1 Å.

The main effort in the X-ray examination of protein fibres came from a single individual, William Astbury, who was at the University of Leeds for 33 years, beginning in 1928, first as lecturer in textile physics, with a decidedly industrial orientation, and later, when his aspirations turned towards molecular biology, as professor of molecular structure. He studied the fibre diffraction patterns of wool, silk, mammalian hair, bird feathers, myosin, epidermin, fibrin, collagen, almost anything he could easily get his hands on. He proposed detailed structures for most of these proteins, and we should state at the outset that he was nearly always wrong in his interpretations. His influence came from his enthusiastic advocacy of X-ray analysis rather than his use of the method to produce the ultimate solution to structural problems.

Astbury (1898–1961) had the best possible credentials as an X-ray crystallographer, having been a member of William Bragg's research staff, first at University College, London, then (after 1923) at the Royal Institution, where Bragg followed in the footsteps of Faraday and many other famous physical scientists as director of the Faraday laboratory. Astbury got into textile analysis more or less accidentally, when Bragg asked him to produce some X-ray photographs of 'common things' for one of the Royal Institution's traditional Christmas lectures for children. This experience made Astbury a logical candidate when the position at Leeds became available. His first project there was to contribute to studies of the elastic properties of wool that had been begun a few years earlier by J. B. Speakman. Astbury set out to investigate hair, wool, and related fibres, for all of which the protein keratin was the principal constituent. It would prove to be effective and influential work—he was the first on the scene and the dramatic facts were there waiting to be discovered.[17] As Speakman's work had indicated, the fibres could be stretched and would

---

measure of particle asymmetry became a familiar parameter for protein characterization. See, for example, the discussion of antibody protein in Chapter 23. But the future lay by common consent with X-ray crystallography.

snap back to their original length when the stretching force was relaxed. The X-ray data for the two states proved to be different, demonstrating for the first time a measurable change in protein structure at the most intimate molecular level—interatomic spacings at just a few ångströms. Moreover, the change from one to the other was reversible. Astbury called the two forms α-keratin and β-keratin: the α/β designation has persisted and has become universally associated with the underlying polypeptide chain conformational change.

As it happened, the X-ray data from wool and hair yielded only blurred diffraction spots, not nearly as sharp as those obtained by Rudolf Brill in Germany a few years earlier from silk, but they were good enough to permit limited quantitative interpretation. The stretched form (β-keratin) was immediately recognized as substantially identical with silk fibroin, with a 3.4-Å repeat in the fibre direction, compatible with a fully stretched polypeptide chain.[18] The unstretched form (α-keratin) had a 5.15-Å repeat in the fibre direction which was also interpretable, but not as a straight chain, only as a shorter, somehow folded chain. Periodicities perpendicular to the fibre axis were observed as well: one at 4.5 Å, seen both in silk and in β-keratin, interpreted as corresponding to the distance between adjacent polypeptide backbones. It was missing in α-keratin, consistent with the notion that polypeptide chains in that state were somehow folded.[19–21]

Astbury's speculations

Astbury speculated rashly about the precise details of the α structure. He noted that there is an important 5.15-Å repeat distance in cellulose fibres as well as in keratin. In cellulose the distance corresponds to the length of a glucose residue in the polymer chain and Astbury suggested that a six-membered ring might likewise be the structural basis in keratin. Diketo-piperazine is a hexagonal ring well-known to peptide chemists. It is a closed ring and therefore cannot be an element of a fibrous chain, but Astbury thought it might serve as a *model for a fold* in a continuous chain, as illustrated by Fig. 6.1. The idea in retrospect is chemically absurd, even if we overlook the fault that the back-and-forth excursions in Astbury's picture are conceived as planar instead of three-dimensional. Linus Pauling, twenty years later, is said to have been shocked by the liberties taken by Astbury and other crystallographers of the time in the devising

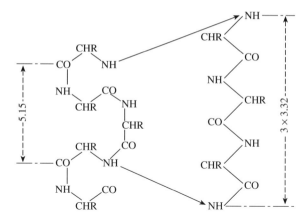

Fig. 6.1 Astbury's speculative picture of the keratin α ⇔ β transition. The numbers in the axial direction represent repeat distances in ångströms. (Based on Astbury and Street, 1930.[20])

of molecular models. They seemed to think (according to Pauling) that if the atoms were arranged in the right order and about the right distance apart, that was all that mattered. Retrospectively, this seems to be a case of the pot calling the kettle black. Pauling himself, though he was right about polypeptide chain configurations, was often rash and wrong about other structures—most notably that of DNA, as is vividly recounted by James Watson in his book, *The double helix*.[22] As for Astbury, he altered his original model when it was pointed out to him by Hans Neurath in 1940 that there was no room for the side chains in his proposed keratin structure.[23]

Astbury went on to study other protein fibres.[24] Feather keratins turned out to be much more complex than hair and Astbury suggested another wildly speculative molecular structure, containing *lateral* polypeptide chains grafted onto the main chain through peptide bonds from side-chain carboxyl or arginyl residues. Muscle fibres and the blood-clotting protein fibrinogen, on the other hand, were grossly similar to hair and wool and presumably could also exist in α or β conformations.[25] At the opposite extreme was collagen, clearly a radically different kind of fibre, with no resemblance to either α or β. Astbury noted its unique amino acid composition, dominated by proline and glycine residues, but made no specific structural suggestion.

Astbury was a popular lecturer at conferences and was valued for his often stimulating and productive conceptual contributions. Thinking about the rich content of the double-headed amino acid cystine of both wool and hair, he suggested that cystine might form cross-links *between chains*—that is, half of each cystine residue could be in one chain, the other half in another. Also, he was fascinated by newly published X-ray data for pepsin, indicating far more internal order than previous data for three-dimensional protein crystals had shown.[26] His intuitive thought[27] was that fibrous structures might be the basis for the crystallinity: globular proteins in general might be folded from elements essentially like the elements of fibrous proteins! How right he proved to be, but that was much later. Bernal in his biography of Astbury considered this recognition that there might be no radical difference between fibrous and crystalline proteins as one of his great contributions.[17] Overall, in Bernal's words, Astbury 'influenced everybody's thinking about large biological molecules' and was 'the father of all those who since then interpreted other types of fibrous structure . . . and who can recognize types of twist from the pattern of blurs on rather obscure fields'.

CHAPTER 7

Analytical imperative

How can it happen, one may ask, that something apparently so commonplace as a separation method should be rewarded by a Nobel Prize?

A. Tiselius, 1952[1]

Primitive classification

Even as early as 1905, when Emil Fischer and Franz Hofmeister had only recently established the nature of the peptide linkage, the number of proteins recognized as distinct in terms of either function or physical characteristics was becoming uncomfortably large. How were these increasingly numerous substances to be classified and named? The matter became sufficiently important that two separate committees were set up—one in England consisting of physiologists and chemists and one in the United States, sponsored jointly by the American Society of Biological Chemists and the American Physiological Society. Their reports appeared in 1907 and 1908, respectively.[2,3]

The two reports were similar in substance. They regretted the lack of adequate chemical knowledge to propose a new approach to the problem and felt they had to content themselves with a formalization of existing popular nomenclature and elimination of ambiguities that had cropped up within it. For *simple proteins* the only available systematic nomenclature was based entirely on solubility: *albumins*, for example, were defined as soluble in pure water; *globulins* were insoluble in water, but soluble in neutral salt solutions; *glutelins* required dilute acid or base for solubility; *histones* (recognized as composed mainly of basic amino acids) were very soluble in water, but insoluble in dilute ammonia. Coagulability by heat was part of all definitions, deemed to be a common property of all simple proteins. The committees went on to define conjugated proteins of various kinds, such as haem proteins and nucleoproteins, and gave names to fragments obtained by proteolysis, recognizing primary and secondary products of hydrolysis as distinct—the macromolecular nature of proteins was all but universally accepted, as we already stated in Chapter 4.[4]

Both reports recommended that the term 'proteid', which had often been used interchangeably with 'protein', should be abandoned. Apart from that one beneficial change, the overall effect of the classification scheme on the naming of proteins was dull and unproductive: 'lactoglobulin' gave no information other than to identify a protein as a globulin from milk; likewise, 'serum globulin' was a globulin from blood serum; no similarity to one another apart from solubility was implied.

In spite of its deficiencies, the solubility-based system was retained for many years. Even when improved fractionation methods, such as the use of added salts at successively higher concentrations, led to a need for refinement, the resulting modifications tended to stay within the old scheme: names such as 'euglobulin' and 'pseudoglobulin', for example, were assigned to fractions of blood serum globulin. But as the pace of research accelerated and new proteins were being discovered almost every day, a dramatic change became necessary, not just in classification and nomenclature, but in the actual laboratory methods on which classification and nomenclature are based. More versatile methods were needed for separating the components of a protein mixture and for chemical characterization of the separated components. As regards the latter, what had to be faced was the reality that the only acceptable *initial* (empirical) characterization was by amino acid composition—exact numbers for every amino acid residue in every known polypeptide chain. Simple organic compounds were customarily first defined empirically on the basis of atomic composition—C, H, O, N, plus other elements if present. For proteins this was inadequate: amino acids were the only logical basis for empirical formulas. And because proteins were not necessarily abundant in the natural sources from which they were derived, speed of analysis and minimal sample size were part of the problem that had to be tackled.

This chapter will give an account of how these goals were actually attained and it may come as a surprise to readers who believe that the flagships of progress in science come inevitably in the form of new ideas, new hypotheses. The purely technical advances described here have played just as vital a role. Brilliant minds were, of course, among those involved, but it must be emphasized that their brilliance was not directed at questions in the 'what is life?' category. They knew only in a general way how new techniques might be applied and their own projects were designed without reference to the kind of central questions or prophecies that brought fame to those who are best remembered by the general public.

There was a corresponding change in spirit in research reports and published papers: nuts and bolts and leaky junctions dominated; there was little time to speculate about how results might be interpreted. The following is an illustrative quotation from A. J. P. Martin, dating back to 1941, when Martin and Synge were developing the use of partition chromatography to separate and measure the amino acids in a wool hydrolysate.

> It was a fiendish piece of apparatus, we had to sit by it for a week for one separation; it had 39 theoretical plates and filled the room with chloroform vapour. We used to watch it in 4-hour shifts. We had constantly to adjust small silver baffles to keep the apparatus working properly.[5]

Separation was a prerequisite for analysis and, for proteins, old chemical methods were inadequate: new approaches were needed.

Electrophoretic resolution

Arne Tiselius (1902–1971) provided the spark that set the revolution in motion. He came from a scholarly family and was encouraged to follow his own interests and intellectual fancies in the choice of an academic career; he made the inspired choice to study with The Svedberg at the University of Uppsala. Unlike Svedberg, who began his career encumbered by the unproductive visions of colloid association theories, Tiselius held right from the start an essentially modern molecular image of proteins—engendered in part by Svedberg himself and his ultracentrifugal work. His declared ambition was to find methods to separate the abundant proteins of biological sources from each other, to see them as distinct individuals, where as yet unpredictable insights into molecular mechanisms beckoned.[6,7]

The ultracentrifuge itself had potential as the basis for a method of mass separation, but Svedberg was reluctant to use it that way, having visualized it when he began as a sophisticated analytical tool, designed to yield quantitative data on the compositions of mixtures. The mixtures were for a long time expected to have continuous distributions of mass and mathematical analysis thereof might reveal the key to the mysteries of colloidal aggregation. We have Svedberg's own words on the subject, replying to the director of the Lister Institute in England, who wanted to have a preparative ultracentrifuge designed:

> The types of ultracentrifuge which we have worked out here are not meant for the separation 'in substance' but for optical observations. It has taken us many years of hard work to arrive at the present types, and it may very well take much time and labour to modify the machines for the purpose you indicate.[8]

Electrical force provided an alternative to centrifugal force for generation of molecular motion, possibly more easily adapted to separation, and Svedberg suggested it as a thesis project for Tiselius: to develop the theory and hardware for the study of electrophoresis, a kind of follow-up of Svedberg's own work. The project led to a Ph.D. in 1930 and soon thereafter to a year's Fellowship in the United States, to study with H. S. Taylor at Princeton University, in order to improve his understanding of the physical chemistry of *adsorption*. Later on (in 1948), when Tiselius received a Nobel Prize for his work, both electrophoresis and analysis by adsorption were mentioned in the citation.

At Princeton, Tiselius became acquainted with the Princeton branch of the Rockefeller Institute, which included such luminaries as J. Northrop and Wendell Stanley. Their preoccupation with proteins, enzymes, and viruses turned him on to thinking about biochemical problems as such and he became partially side-tracked from his heretofore purely physico-chemical pursuit of his subject. When he returned to Sweden he radically redesigned his electrophoretic apparatus and put it to its first purposeful use: the analysis of the proteins of blood serum.[10] The result was dramatic. Using horse serum, he found that it could be separated into four well-defined principal fractions: albumin and three globulins, which he named α, β, and γ. All four moved as anions at pH 8. Isoelectric points of the individual fractions, where the direction of movement became reversed, were respectively at pH 4.64, 5.06, 5.12, and 6.0. In describing his results, Tiselius noted that antibody function migrated only with the gamma globulin fraction; the α fraction had the highest concentration of what was still called pseudo-globulin.[11]

It was the identification of immunological activity in the γ fraction that created instant world-wide fame for the electrophoretic method. Tiselius included as part of his first published results an experiment that has since come to be recognized as the classic experiment on the subject. He

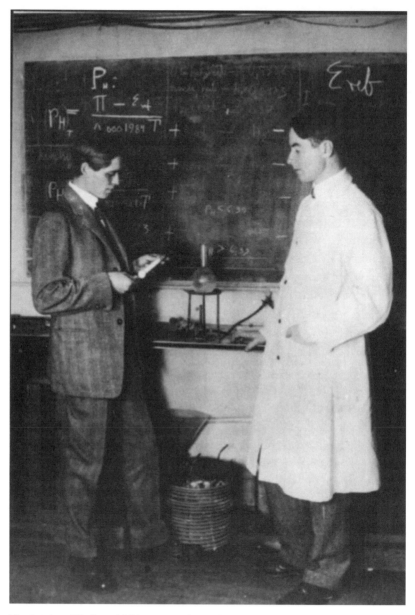

Fig. 7.1 Arne Tiselius with his mentor The Svedberg. This picture was taken in 1926 by Kai Pedersen, long-time associate of Svedberg and biographer of both men. (Ref. 7, courtesy of G. Semenza.)

analysed serum obtained from a single normal rabbit and repeated the analysis on the same animal after it had been immunized against crystalline egg albumin. He found a huge increase in the gamma-globulin fraction. Moreover, this fraction was now 85 per cent precipitable with egg albumin.

Tiselius's electrophoretic method is called *boundary electrophoresis* and is similar in principle to the method used by Picton and Linder in 1892,[12] when they reported the movement of haemoglobin molecules in an electric field, back in the early days when proteins were first unambiguously recognized to be true molecules, as distinct from colloidal aggregates. Tiselius, seeking quantitative measurement of electrophoretic mobilities, had of course put in a great deal of work on the design of his apparatus—he needed to avoid convection currents, to find the best optical systems for following the progress of boundary movement, etc.[13] Picton and Linder's detection of boundary movement had been made possible only by the fact that haemoglobin is coloured; blood serum is colourless and refractive index had to be the basis for optical detection, as shown, for example, in Fig. 7.2.

Many technical advances have followed in the years since Tiselius. Free boundary electrophoresis is no longer the method of choice, at least not when analysis is the goal, and has been largely replaced by *zone*

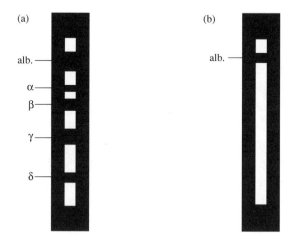

Fig. 7.2 Electrophoretic separation of the components of serum globulin. Part *a* represents whole serum; part *b* shows that the albumin fraction isolated from whole serum behaves as a single component. This is the first introduction of the α, β, and γ notation to designate the individual globulins. (Based on ref. 11, 1937.)

electrophoresis, first on paper, then on starch blocks, most recently on polyacrylamide gels. Various modifications are now available that combine electrophoresis with chromatography. Different solvents can be employed to effect separation of proteins or peptides, with a huge range of possible purposes in mind: for example, the detergent SDS under appropriate conditions separates individual polypeptides rather than whole native proteins. This and other techniques are standard tools in today's laboratories and doubtless familiar to most readers.

But none of the fancy new devices, however remarkable for their speed, resolution, and brilliant versatility, carries quite the allure of that single word, 'gamma globulin', which became a certified word of the English language, one of the few scientific terms whose origins the general public understood and appreciated.

Partition chromatography

We have credited Tiselius with triggering the revolution that altered the face of protein analysis in 1937. From a purely scientific standpoint, it might actually have been triggered thirty years earlier, when the Russian botanist Michail S. Tswett (1872–1919) invented chromatography, doing so in the course of demonstrating that the plant pigment chlorophyll could be readily separated into several components. Specifically, he found that chlorophyll could be selectively adsorbed to paper or other adsorbents and that the separate components then became visible as coloured circles (when allowed to spread on paper) or as rings (when eluted from a column). Tswett later named the method 'chromatography' because visible colour provided the basis for detection. Tswett's work was not appreciated in his lifetime, but it was not for lack of trying, for he was a forceful advocate of his own efforts and did not hesitate in his publications to criticize others, whose data or interpretations (with respect to the chemistry of chlorophyll) he considered erroneous. Retrospectively he is seen to have been vindicated in most such controversies.

Reasons given for the general neglect of Tswett's work include the fact of its novelty (it was evidently 'outside the border lines within which chemists were accustomed to work')[14] and the fact that Tswett never achieved career stability, being forced to move frequently from place to place, often as a fugitive from German troops.[15] Tswett had published some of his work in German journals and had demonstrated his new technique at a meeting of the Deutsche Botanische Gesellschaft on 28 June 1907,[16,17] but the work was not given the weight that it would have had if done in Germany itself. Later on, World War I prevented personal contacts with knowledgeable and potentially powerful German audiences.

By the 1930s the power of column chromatography became widely appreciated, first for use with additional plant pigments and then for other organic mixtures. In relation to proteins, the work of A. J. P. Martin (1910–..) and R. L. M. Synge(1914–1994) would prove to be the most important. This work actually began when Martin was at Cambridge, working on vitamin E, but before long (1938) they were both at the Wool Research Laboratory in Leeds and were collaborating on a project to separate the amino acids obtained from a hydrolysate of wool. Their efforts led eventually to Martin and Synge being awarded the Nobel Prize for Chemistry in 1952. The presentation speech, given by none other than A. Tiselius, begins by asking: 'How can it happen, one may ask, that something apparently so commonplace as a separation method should be rewarded by a Nobel Prize?' Tiselius went on to answer the question himself in a lucid history of the subject; this is a rare occasion when the presenter was himself a world authority on the subject and a previous Nobel laureate (1948).

Martin and Synge made major advances on both theoretical and practical levels.[18,19] They recognized the intrinsic similarity between chromatography and the separation of volatile substances by distillation—in both cases repeated partition between two phases is involved and both are capable of very exact thermodynamic description. The first practical apparatus to exploit this principle consisted in fact of a sequential chain of separatory funnels by means of which a solute could be repeatedly partitioned between two immiscible solvents moving in opposite directions. Needless to say, the necessary apparatus was extremely cumbersome. Apparatus became greatly simplified (without loss of theoretical rigour) when Martin and Synge realized that only one liquid need actually be moving; the other could be trapped by a solid support, such as silica gel—making the apparatus resemble that used for chromatography. But it involved an important modification of conventional *adsorption* chromatography. The solid support was no longer a conventional solid adsorbent, where specific chemical binding affinity is the criterion for selective distribution, but it had instead become an inert material, serving only as mechanical support for a liquid phase. To quote Martin and Synge directly: 'partition between the two liquid phases replaces adsorption on an adsorbent'.[20] The technique became known as *partition chromatography* in order to make the distinction.

Eventually, filter paper was found to be even neater than silica gel—water now held by capillary action to the paper, with continuous re-equilibration as butanol flows over the paper. Each new technical development resulted in better separation with ever-decreasing quantities. 'Amino acid analysis before we began our work required half a kg of protein; ... the silica partition columns require a few mg and paper chromatograms only a few micrograms.'[21]

As everybody knows, the subsequent development of chromatography has been explosive; it has become a mainstay of analysis at all levels of protein chemistry. Martin and Synge were separating amino acids from protein hydrolysates, but new versions of chromatography became applicable to proteins themselves or to peptide fragments derived from incomplete hydrolysis. Here partition between two solvents on the basis of chemical affinities was no longer the ideal tool, and molecular charge or size got into the picture instead. Given that ionic charges are such a prominent feature of proteins, ion exchange chromatography was an obvious first choice.[22] More innovative was the development of gel filtration (also called sieving chromatography): in this technique the stationary phase is a resin that has the important characteristic of having 'holes' of molecular dimensions. Residence time in a hole depends on how big a particle is.[24–26]

Two-dimensional chromatography on paper or flat gels soon followed, using a second eluting solvent in a direction perpendicular to the first, or chromatography in one direction and electrophoresis in the other.

Amino acid analysis

Quantitative total amino acid analysis was universally perceived as an essential objective of protein chemistry. It would do for proteins what elemental composition did for simple organic substances—provide an empirical formula and a minimal molecular weight. William Stein (1911–1980) and Stanford Moore (1913–1982), who set out to attain this goal, cite Max Bergmann at the Rockefeller Institute as the most insistent champion of analysis and the person who inspired them in particular, but an unbiased view of the literature indicates he was only one of many champions. Bergmann, who had come from Europe as an investigator at the Rockefeller Institute in 1934, had a unique motivation—visions beyond analysis itself, seeing possibilities of regularities that might unlock untold secrets of protein chemistry or biosynthesis.[27,28] But his predictions were invalid and he had no direct influence on technical progress: his inspiration of Stein and Moore would be his major contribution.

World War II interrupted the pursuit of protein chemistry at the Rockefeller Institute. Max Bergmann died in the midst of it and Stein and Moore (still inexperienced youngsters) were diverted to more pressing projects. In other parts of the United States, however, the pace of research into one particular group of proteins (the proteins of human blood) accelerated, under a huge government project to fractionate and partially characterize these proteins for the purpose of military medicine, aiming to reduce dependence on transfusions with whole blood. A progress report, comprising 23 papers from several different universities, published in July 1944 in the *Journal of Clinical Investigation*, testifies to the ambitious scope of the project.[29] A later summary, concentrating on the purely scientific, non-clinical aspects of these government-sponsored projects, emerged from a discussion meeting of the New York Academy of Science in October 1945.[30] Fractionation *per se* was in the hands of the Harvard-based group of Edwin Cohn and John Edsall.[31]

Erwin Brand at Columbia University was in charge of the analytical work and at the New York meeting he presented a comprehensive review of all available results for amino acid content.[32] One small but useful contribution was to propose a simple system of symbols. The first three letters of the name of each amino acid were to be used whenever feasible; the few exceptions were Ileu for isoleucine, Asp-NH$_2$ and Glu-NH$_2$ for asparagine and glutamine, CySH and Cys-, respectively, for reduced and oxidized cysteine. These symbols were instantly endorsed by everyone and were retained for many years with only minor modifications—for example, Asn and Gln for the acid amides. Many people did in fact report empirical formulas using these symbols, for example, Gly$_{27}$Val$_{34}$Leu$_{46}$ . . . , etc., in analogy with atomic formulas for simpler organic molecules.[33]

The most striking aspect of Brand's review is in the analytical methods that were being used, presumably representing the state of the art at the time the project was planned, fixed in advance, made resistant to change by the urgency of war and the inflexibility of government budgets. We see almost total reliance on old methods, judged in retrospect to have been on the verge of obsolescence even then. There was no experimenting with the use of chromatography. A few results reported by Martin and Synge were included in Brand's review, but they were at that time limited to amino acids with non-polar side chains, the reason being that N-acyl derivatives of the amino acids were employed, which converted the normally charged amino group into an uncharged entity (poorly soluble in water) and required an organic solvent for elution. Martin and Synge, who were known at this critical period (if at all) only for their analysis of wool,[34] were, of course, not themselves involved in the massive American project centred on blood proteins.

Chemical methods of analysis were practical for a few amino acids, but they were in the minority.[35] For most of the amino acids the best available procedures relied on incredibly cumbersome microbiological assays![36] These assays used mutants of micro-organisms (*Neurospora*, *Lactobacillus*, etc.) that lacked the ability to synthesize a particular amino acid. Growth of cultures was then quantitatively dependent on the amount of that amino acid in the culture medium and, after proper calibration, could provide quantitative analysis for that particular amino acid. In the most detailed analysis reported up to that time for any one protein (β-lactoglobulin), about half the data came from microbiological assays, each individual determination requiring day-long incubation with its own specific mutant organism. An untold number of Petri dishes must have been involved.[37]

The most authoritative voices were full of enthusiasm for the microbiological methods. John Edsall, for example, reviewed recent progress at the New York meeting and singled out

> the introduction of the microbiological methods, which have furnished such rapid and astonishingly accurate estimations of small amounts of many amino acids, including several that were previously quite inaccessible to accurate estimation. The future significance of partition chromatography and allied techniques may be very great indeed, but, at present, the figures derived from these methods appear to be subject to somewhat greater error than those derived from some of the other methods under discussion.[38]

Even H. T. Clarke, a physical chemist who might have been especially prejudiced against methods based on the use of live organisms, went out of his way to praise microbiological methods in his concluding comments for the New York conference: 'Precision is astonishingly great . . . . A brilliant future can safely be predicted for these methods.'[39]

William Stein and Stanford Moore would prove otherwise. They resumed their collaboration at the Rockefeller Institute after the war and in the interim had been completely convinced of the superior prospects for partition chromatography. They worked with unmodified amino acids in place of the N-acyl derivatives used by Martin and Synge, a great improvement because it avoided the need for organic solvents in the separation. They employed starch columns instead of silica gels as stationary phase; they rediscovered an ancient colour reagent (ninhydrin), which enabled them, after proper calibration, to quantitate the amount of amino acid in each fraction.[40] The method was perfected in 1949 and it was used for a complete analysis of the two proteins, β-lactoglobulin and bovine serum albumin, for which the results of Brand's group were deemed to have the been the most reliable. Results were most gratifying.[41] It is noteworthy that Brand had proudly proclaimed his β-lactoglobulin paper as 'the first to describe a complete analysis of a single preparation of protein by the same group of workers', a statement that vividly illustrates the obvious difficulties that existed when each amino acid in a protein hydrolysate required a separate method of analysis (for example, its own specifically designed micro-organism).[42] It had taken only three short years, from 1946 to 1949, for chromatography to take over—like a breath of fresh air, one might say.

Starch columns were subsequently deemed to be too slow and were replaced by ion exchange resins. Also, Stein and Moore developed the mechanical devices now in universal use, the drop counter and the fraction collector. With the help of commercial laboratories, it was all put together for automated performance. By 1958 an automatic analyser could do its job overnight, without the need for an operator in attendance.

CHAPTER 8

Amino acid sequence

The enzymatic breakdown of proteins always leads to mixtures of many substances of unequal molecular size . . . . The isolation of definite chemically-characterized albumoses or peptones from these mixtures is a thankless task.

F. Hofmeister(1908)[1]

Knowing the amino acid composition of a protein hydrolysate, even with high accuracy, is just like having a sackful of letters of the alphabet, all mixed up. Knowing the arrangement of these amino acids along a polypeptide chain—or, sometimes, several connected polypeptide chains —is another matter altogether: words and sentences are now formed. This experimental transition was first achieved by Cambridge biochemist Frederick Sanger (1918–..) during a period lasting more than ten years, from about 1945 to 1955.

The amino acid sequence of beef insulin

The beginning of the project may be marked by a curious feature of Erwin Brand's amino acid analyses. He was the great authority on overall amino acid composition, to whose work we referred in the preceding chapter. He especially favoured the use of microbial assays, each specific for a single amino acid, but his data invariably included a figure for free α-amino groups—that is, amino groups at the ends of polypeptide chains, not involved in peptide links. These figures could not, of course, be measured by microbiological methods, but were based instead on combining a direct chemical procedure for the *total* content of free amino groups, as measured before proteolysis, with the specific value in the hydrolysate for lysine, the one amino acid known to have a free amino group on its side chain. Subtracting one from the other should yield the desired number of chain-terminal α-amino groups. This procedure, involving subtraction of one large number from another, is subject to huge experimental error, but Brand did not seem to care and equated his result uncritically with the number of independent poly-peptide chains in the protein he started with. His data indicated the presence of as many as ten or more independent polypeptide chains in common proteins, such as the serum albumins, γ-globulin, and insulin. Other more direct ways of measuring the number of polypeptide chains, not involving organic group analysis, disagreed with Brand's estimates, but this did not mean that the latter could be dismissed out of hand.

Sanger, an exceptionally modest man, explains how he got involved.[2] He 'happened to get a job' with A. C. Chibnall at Cambridge and was set the problem of devising a truly reliable analytical method that would be specific for α-amino groups.

Chibnall had had a year of postdoctoral study with Thomas Osborne in his formative years and had been in the amino acid analysis business since 1933, using traditional methods of organic chemistry, where each individual amino acid had to be isolated separately from hydrolysates before its quantity was measured. Brand's exhaustive paper on the amino acid composition of β-lactoglobulin[3] included 'literature' values to compare with his own and one of them was a value for arginine that had been determined by Chibnall.

The result of Sanger's specifically focused efforts was a new general method for chain-terminating α-amino groups based on the reagent dinitrofluorobenzene.[4] This compound formed bright yellow dinitro-phenyl (DNP) derivatives with the terminal group, and did so without breaking peptide bonds. Then, when peptide bonds were cleaved by hydrolysis, the DNP linkage usually remained intact. The terminal amino acid could be spotted in a chromatogram by its yellow colour and identi-fied by ordinary procedures. However, one of the yellow derivatives, DNP-glycine, turned out not to be as stable as the rest and required a shorter proteolysis time if it was to remain detectable, which led to the need to compare short and normal proteolysis times routinely, to be sure nothing had been missed. Such comparison, however, was found to yield some extra yellow spots, that originated from dipeptides (and sometimes other short peptides) that still had an intact peptide bond between the terminal amino acid and the one that followed it in the polypeptide chain. Such peptides could then be isolated and the amino acid adjacent to the terminal one could be identified.

Serendipity! It became obvious that sequence determination was in principle possible by following up this result.

An enormous amount of work was, of course, required to turn what was in principle possible into accomplished reality. For example, degradation methods had to be studied in detail, using acid, base, proteolytic enzymes, and other means, to make sure that they worked quantitatively and that no rearrangement occurred during analysis. With this information, it was no longer necessary to limit oneself in sequence analysis to intact entire polypeptide chains: fragments could be obtained by partial hydrolysis and could be sequenced separately, overlaps in short lengths being used to fit it all together. It was also desirable to explore methods for sequencing from the other end of a polypeptide chain, the terminus with a free carboxyl group. The proteolytic enzyme carboxypeptidase turned out to be capable of this task, being specific for the C-terminal peptide bond. Its rate of action was found to depend significantly on the identity of the terminal amino acid, making approach from that end trickier to use than the N-terminal approach, but it was nevertheless a valuable tool.

Sanger concentrated his work on a single protein, insulin, one of the smallest known. Its molecular weight from physical measurements was 12 000 and on this basis it was expected to have 102 amino acid residues per molecule. Sanger identified four terminal amino groups per molecule, indicating the presence of four separate polypeptide chains, two with chain-terminal phenylalanine and two with chain-terminal glycine.[5]

It was then found that these polypeptide chains could not be separated by purely physical means because they were held together by disulphide bonds. This engendered more systematic exploratory work, to learn how best to break disulphide bonds. Oxidation turned out to be the method of choice, with optimal recovery of the now separable Phe and Gly chains.[6]

The separated chains could now be used as sources for N-terminal peptides and it emerged that there were only two distinct chains, not four, suggesting that the molecular weight of 12000 must represent a dimer.[7] The two chains comprised a total of 51 amino acid residues, 21 in the Gly-terminal chain (now named the A chain) and 30 in the Phe-terminal chain (B chain). The first total sequence to be completed was that of the B chain, published in 1951.[8,9] The A chain followed two years later.[10] About the same time, independent new data from other laboratories

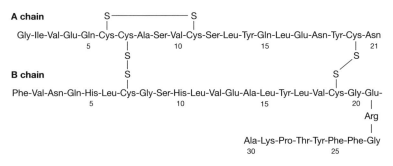

Fig. 8.1 Complete amino acid sequence of beef insulin. (Based on ref. 11.)

confirmed the smaller size for the whole molecule. Sanger's actual numerical value for the overall molecular weight, based on the sums of the weights of the amino acid residues, was 5734.

At this point the separate chain sequences were known, but it was not enough; more arduous work was in store, for the separated chains had cystine in the broken oxidized state, at four locations in the A chain and two in the B chain. There were many ways in principle to pair them off, forming three disulphide bonds. To show the original positions of the disulphide bonds in the native protein required going back to the beginning and carrying out partial hydrolyses by various methods under conditions where disulphide bonds remained intact, followed by separation and purification of the resulting peptides, and (finally) sequencing of the latter to fit them into the already established sequences for the A and B chains. Care had to be taken to avoid artefacts due to disulphide interchange. It took another year to sort this all out and the complete structure was published in 1955 and is shown in Fig. 8.1.[11]

From start to finish, the project required more than 10 years, a monumental tribute to patience and persistence. It must be appreciated that this was uncharted territory, fundamental organic chemistry at a level not attempted since the days of Emil Fischer and his disciples. Analytical tools to separate and recognize reaction products had become highly sophisticated, but the reactions themselves were mostly original. Sanger did nearly all the work by himself, in a small laboratory, with only a handful of assistants. A Nobel Prize duly followed in 1958; Sanger went on to merit a second such award in 1980, for developing methods for sequencing nucleic acids.

Fig. 8.2 Fred Sanger in the laboratory. (Photograph kindly provided by Dr Fred Sanger.)

Other species

All these studies had been done with beef insulin, but the problems were now ironed out and, with the aid of some undergraduate students, pig and sheep insulin were also sequenced quickly. The Phe chain proved to be identical in all three insulins; the Gly chains were all the same length, but differed by substitutions in positions 8 to 10.

It is perhaps the best possible illustration of the huge step forward in protein chemistry that was achieved by the insulin sequence—that we can almost casually, in two sentences, describe precisely how an important protein differs between one animal species and another.[12]

Original sceptics

To broaden the historical perspective, it is desirable to note that published comment on the sequencing work, as it was in progress, demonstrates a certain amount of scepticism about the very idea of there even being a unique sequence. This was a consequence of the fact that many proteins other than insulin were known to have not only variations between species, as observed in insulin, but also presumably genetic variations between individuals within the same species. They were recognized electrophoretically, on the basis of differences in overall electrostatic charge, the most famous being the distinction between normal and sickle-cell human haemoglobins,[13] which was discovered in 1949. It was possible that a multitude of such variants could exist, but remain unrecognized because no change in electrostatic charge was involved.[14] The word 'microheterogeneity' was used to describe this situation. A broad review of the subject was published as late as 1954, with the conclusion that 'all protein preparations prepared so far represent populations of closely related members of a family, not collections of identical molecules'.[15] This position could be speculatively supported even after Sanger's work was published because his sequence might simply have represented the dominant member of a mixture. But, if so, Sanger would have been expected to find some anomalies when he was aligning proteolytic fragments in insulin on the basis of overlaps within their sequences—no such result was observed at any significant level.

Another original sceptic had been the French molecular biologist Jacques Monod. As cited in a retrospective interview by H. F. Judson, he assigned a sweeping historical significance to Sanger's sequence determination: he is quoted as saying that one 'could not even have begun to think seriously about the genetic code' until after Sanger's work. According to this account, many biochemists still believed (before Sanger) that the Bergmann–Niemann numerology for amino acid content or something like it would prove to be valid, indicating that polypeptide chains were assembled by some as yet undiscovered chemical rule from short repetitive sequences—in which case 'you didn't need a genetic code'.[16] It is difficult to judge the merits of verbal statements such as this, made twenty years after the facts. It may well have been true with respect to the effect on Monod's own work, but the generalization to others is questionable. Francis Crick, in his own retrospective book, has a more credible assessment. Before Sanger, he says, iron-clad ordering of amino acids could not be established beyond a shadow of a doubt, 'but it was easy enough to guess that this was likely to be true'.[17]

CHAPTER 9

Subunits and domains

The concept of subunits, more than one polypeptide chain in a protein molecule, emerges automatically from investigation of amino acid sequences. We saw that in the preceding chapter: the very first complete sequence analysis, that of insulin, began with an accurate determination of free α-amino groups and showed that insulin belongs in the multi-subunit category. (This particular example will be seen later in this chapter to be more complex than at first it seemed.)

Another concept that emerged from investigation of amino acid sequences was the concept of domains within a single chain, chemically and/or functionally distinct subdivisions that may become apparent only after a sequence has been completed. Insulin does not provide an example, but many occurrences (with quite diverse origins and interpretations) were found as amino acid sequence analyses began to be applied systematically.

Haemoglobin subunits

Given that we are writing a history, we must note that the concept of subunits is actually much older than the art of sequence determination. Haemoglobin is the most famous example: it has been known since 1924, one of the most firmly established facts in protein chemistry, that a haemoglobin molecule is composed of four subunits. The mass per haem group or per iron atom had consistently been measured for fifty years or more as around 17 000. The overall molecular mass, on the other hand, was clearly four times greater. This was established independently and essentially simultaneously in Cambridge and in Uppsala, by people with quite different motivations.

Cambridge was the world centre for haemoglobin research, where the celebrated blood physiologist Joseph Barcroft worked and aroused the interest of many of his colleagues in the haemoglobin molecule as a chemical and functional entity. Among them was the biochemist G. S.

Adair, who set out to measure the molecular weight by osmotic pressure. He went to exhaustive lengths to identify and correct for all conceivable errors in the measurement, which originated principally from movement of small ions across the protein-impermeable membrane. He measured a molecular weight of around 65 000 both for haemoglobin dissolved at high salt concentration and for haemoglobin dissolved in salt-free water, where the impact of just a few extraneous ions could become disproportionately important. It was an experimental triumph, recalled in the memoirs of many who visited Cambridge at the time.[1]

These results were supplemented by measurement of the equilibria for oxygen binding as a function of oxygen pressure, which, if they were to be explained by proper thermodynamic analysis, proved to require 'cooperativity' between several binding sites (more than two) on the same molecule. The equations that resulted from this work formed the basis for theoretical models that would prove to be key factors in understanding the details of the physiological function of haemoglobin and, subsequently, many enzymes.[2–6]

In Sweden in the same year, The Svedberg, in almost his first use of the ultracentrifuge as an analytical tool for proteins, confirmed the molecular weight of 67 000, independently of the Cambridge group—not even knowing at the time of Adair's osmotic pressure results. Pedersen tells a revealing anecdote about this.[7] Svedberg, just recently converted from the ideas of loose colloidal association to the conviction that proteins were proper macromolecules with fixed molecular weight, firmly believed that the molecular weight would be around 17 000 and in that case nothing exciting would come out of the first haemoglobin run on the ultracentrifuge—the centrifugal force being used was too weak to produce a sharp sedimentation boundary for a particle of that size. Svedberg went home and left his colleague R. Fåhreaus in charge of the apparatus overnight. But, the molecular weight being four times larger than expected, the sharp boundary duly appeared. Fåhreaus was so excited that he woke Svedberg in the middle of the night with a phone call. 'The, I see the dawn,' he is quoted as saying, as a clear space developed at the top of the centrifuge cell above the sedimenting coloured protein.

In the modern era, with sequencing and advanced analytical tools at our service, we have learned that many proteins do consist of a single polypeptide chain per molecule, but that proteins formed by association of several separate chains are equally common. Some oligomeric proteins are like haemoglobin, in that their constituent polypeptides are similar or even identical. Others are more complex: the ATP synthase that is central to mitochondial and bacterial energetics is an assembly of 22 subunits of 8 distinct kinds, ranging in individual molecular weight from 8000 to 55 000.[8]

Sometimes, as in haemoglobin and in ATP synthase, subunits are held together purely by non-covalent forces (but nevertheless strongly and specifically) and can be separated without breaking chemical bonds. In other cases disulphide bonds covalently link one chain to another—as shown for native insulin in Fig. 8.1 of the preceding chapter, for example, and as is also true for the common globulins involved in the immune response. When this is observed, the disulphide linkage is always, as in insulin, between uniquely specified cystines in the chains being joined.

Precursors

Insulin actually happens to be an example of a protein that is altered after it is first synthesized.[9] This in itself is a not infrequent occurrence, but in the case of insulin it happens to change the polypeptide count. The precursor, proinsulin, is a *single polypeptide chain*, with intrachain disulphide bonds.[10,11] It is attacked by proteolytic enzymes on its passage through

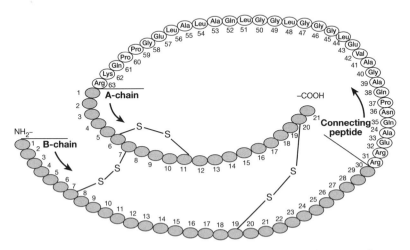

Fig. 9.1 The primary structure of the precursor protein, proinsulin. The structural diagram includes definition of the connecting peptide between what would ultimately be separate chains. (Based on the work of Donald Steiner and co-workers.[10,11])

the pancreatic duct, leaving the structure determined by Fred Sanger. The latter is by definition the *native* structure, as the term is normally used to designate a protein in its physiologically active state, not for the protein emerging freshly from its point of synthesis.

Some precursor proteins had been known as such for a long time, before amino acid sequencing could reveal precise details. Fibrinogen, for example, as its name indicates, is the precursor that generates fibrin clots, a relationship already established around 1850.[12] Proteolytic enzymes secreted from the pancreatic duct provide another example. It has been known since 1867 that these enzymes are inactive in freshly prepared pancreatic tissue or juice: their precursors (pepsinogen, chymotrypsinogen, etc.) became well known and were even obtained in crystalline form in 1934.[13] In modern times, precursors have proliferated as biochemists have become more knowledgeable about the early stages in the existence of proteins, from synthesis to their travel to final destinations.

Polypeptide domains

Mention should be made, finally, of the existence of 'domains' within polypeptide chains, segments of a chain that differ significantly from the rest. The difference may be in overall composition, as is often observed in membrane proteins, where some chain segments may be manifestly non-polar, suggesting a destined location within the typically anhydrous interior of a cell membrane.[14] Or the difference may be a stretch of amino acid sequence that is strikingly unusual in some other way: one of the polypeptides of complement, for example—a long-known component of the immune system that can parlay antibody–antigen recognition into actual damage (cell lysis) to an invading cell—contains segments of the repetitive sequence –Gly–Gly–Pro– which is ordinarily seen only as a characteristic element of the connective tissue structural protein, collagen.[15]

Alternatively, the distinction may be purely functional, possibly genetic, without any striking anomaly in sequence segments as such: this is the case, for example, in the variable and constant domains of the polypeptide chains of antibody molecules. (See Chapter 16.) The possibilities are endless and all or most of the documentation dates from 1970 or later.

Concluding comment

Our purpose here has been to point to the dramatic change in perspective that followed directly from the determination of the insulin sequence, beyond the dictate that determination of the exact sequence, having been demonstrated as practicable, became an obvious obligation for every protein chemist, no matter what particular protein he was interested in. The very meaning of 'sequence determination' was altered: definition of subunits, domains, and precursors became almost automatically a part of it and provided insights into protein structure, function, and genetics, which have enriched that pageant of protein chemistry in ways that could not have been predicted in advance. It is a tribute to the far-sighted planning and exacting standards that Frederick Sanger set himself that his work has been found so readily applicable to sequences far longer than the two tiny insulin chains to which he devoted himself with so much patience.

PART II

Detailed structure

CHAPTER 10

Early approaches to protein folding

This leaves us with the paradox that the pepsin molecule is both globular and also a real, or potential, polypeptide chain system.

William Astbury, 1934[1]

Defining the problem

The peptide bond alone, serially connecting amino acids to one another, can structurally do no more than produce polypeptide chains: from dimers and trimers upwards, without limit to the ultimate length. Solid evidence from X-ray analysis for such polypeptide chains, extended to considerable length, existed in the case of normally insoluble protein fibres. It was easy to imagine that proteins that were water soluble contain equally long chains, but flexible rather than stiffly extended—fluidity at the molecular level reflecting fluidity in the medium in which the proteins existed.

The facts, as seen around 1930, were, however, quite different. The proteins that were the most popular objects of study in solution were not extended or flexible chains, but rather they were the so-called globular proteins, appearing as tightly packed entities, almost spherical in shape with diameters often ridiculously small in comparison with the length of the continuous polypeptide chain they were known to contain. What kind of molecular organization was involved? And, more importantly, why should a polypeptide chain not be content to meander loosely through the solvent? What driving force demanded a kind of internal collapse?

The challenge posed by these questions was central to protein chemistry for thirty years. It enticed a diverse group of people into protein science, people from different scientific backgrounds, breaking the conventional boundaries of whatever fields they were trained in, and also people of quite different temperaments. A few were methodical and patient, accepting the existence of a multitude of distinct proteins, each perhaps with a unique internal organization, willingly facing the likelihood of years of labour before the answers to the questions posed became apparent. Others sought new general principles, shortcuts to the promised land. This latter group included truly brilliant minds, as well as would-be geniuses who were nothing of the kind. A common bond was a tacitly agreed permissiveness—*carte blanche* for whatever your vision to future progress might be.*

William Astbury, the protein fibre X-ray expert (Chapter 6) and the source of the quotation at the head of this chapter, who travelled from meeting to meeting lecturing about fibrous structures, had a simple view that turned out to be not too far off the mark. He suggested that there might be no radical difference between fibrous and globular proteins, that the fibrous chains might be formed by linking up of originally globular units or, vice versa, that the chain itself might be the true original unit, which, in forming globular particles, 'is afterwards folded up in some neat manner'. He had the good sense not to try to be more specific.[1,2]

Bernal's prophetic vision[3]

X-ray diffraction is a method ideally suited for the analysis of single crystals, with crystalline order in three dimensions. Fibre diagrams, including those studied by Astbury, were an offshoot, something one could use to gain much more limited information, where crystals are not available. But, for many proteins, true crystals were, of course, available, perfect material for exploitation of the X-ray method in its ultimate detail. John Desmond Bernal (1901–1971) knew this and he was an enthusiast from day one. His very first published paper, in 1924, under the tutelage of William Bragg at the Royal Institution, had been on the structure of graphite,[4] as revealed by X-ray analysis and it was followed by a missionary

---

* It is only fair to point out that the state of knowledge at the very core of biological science was still extremely primitive, encouraging uninhibited hypotheses. Nobody had even the faintest understanding of the actual path of biosynthesis of protein molecules; most people still thought that proteins were the carriers of genetic information—not DNA. The first paper that proved it was DNA was only published in 1944 and the 'proof' was regarded with scepticism for several more years. See detailed discussion in Part IV of this book.

zeal for extension of the method to proteins; Bernal is the model for the character Constantine in C. P. Snow's novel *The search*, the one who had great ambitions to make discoveries in protein chemistry.[5]

More than just championing the tool, Bernal put his convictions into practice, and (with his student, Dorothy Crowfoot Hodgkin) became the first to produce a sharp X-ray picture of a crystalline protein—pepsin in 1934, crystallized after the method just devised four years earlier in America by John Northrop. An immediate discovery was the importance of water in the stability of protein structures: evaporation during crystal examination led to rapid loss of sharp detail in the X-ray photograph. Also, the compact globular shape of protein molecules was confirmed, the molecules being separated in the crystal by spaces which contain water. But beyond that it was not possible to go—fulfilment of Bernal's vision had to await technical methodology that would take 25 years to develop. (See Chapter 13.)

In the interim Bernal became much more than a pioneer of X-ray diffraction: he was prominent in every aspect of science that is even peripherally related to proteins and he will appear in every chapter of this section of our book. He thought about 'structure' in liquids long before anyone else and was responsible for the first model for the structure of liquid water, the framework within which the globular proteins exist and function. Uniquely among protein chemists, he recognized the significance of hydrophobicity for protein folding. And he had a unique personal history as well. Nicknamed 'the Sage', he was the epitome of a liberated free-ranging spirit, possessor of a mind that raced unimpeded from subject to subject. 'All riches of thought was open to me', he wrote about his undergraduate days in Cambridge, and 'I have no one supreme interest and am fascinated wherever I look'. He looked not only to science, but also to philosophy and social theories. He was an avid Marxist and supporter of Russian communism. He was a leading anti-war activist in World War II, but nevertheless was selected to advise his government on the conduct of the war, on the staff of Sir John Anderson, British war-time minister of Home Security. Later he became Lord Louis Mountbatten's trusted advisor on technical problems related to the Allied invasion of Normandy[8] and he accompanied Mountbatten to India and Sri Lanka when the theatre of war moved east after the Normandy invasion. Mountbatten became very fond of Bernal and wrote an appreciation of him for inclusion in the biography[3] by Dorothy Crowfoot Hodgkin, who

Fig. 10.1 J. D. Bernal with Lord Mountbatten and others on military service. (Source: Hodgkin's biography of J. D. Bernal, 1980.[3])

(as Dorothy Crowfoot) had been Bernal's most famous student, subsequently winner of a Nobel Prize for her structural work.[9]

Flights of fancy

As we have said, these were early days, too soon for tackling detailed protein structures by X-ray crystallography. Bernal and Dorothy Crowfoot (who became Dorothy Hodgkin in 1937) returned to structures of less complex organic molecules on which Dorothy had been working before the pepsin crystals arrived—hydrocarbons, vitamins, sterols.[10–13] Research on penicillin began in 1942 during the war and the penicillin structure was finally solved in detail four years later.

In the interim, protein science was for more than a decade in the already-mentioned fluid situation, where everyone who wanted to could have their say and many of them did. Some promising leads arose quite naturally from chance discoveries, the most prominent of these being T. Svedberg's proposal (an outcome of his measurements of molecular weight by use of the ultracentrifuge) that all proteins are built up from some common subunit or from a class of subunits with about the same molecular weight. The striking and unexpected similarity in molecular weight and shape for egg albumin and Bence-Jones protein, discovered early in Svedberg's work,[14] of course contributed greatly to Svedberg's enthusiasm, as did the established facts of haemoglobin dissociation, determined by Svedberg and independently by Adair in Cambridge, who had used osmotic pressure for his measurements.[15,16]

Even more intriguing for those with faith in sudden revelation was the hypothesis of Bergmann and Niemann (1937), claimed to be based in part on Svedberg's proposal and on amino acid content data.[18] Looked at soberly, it proved to be nothing more than numerology, with only the flimsiest supporting evidence. Its precise thesis was that the total number of residues of amino acids in any protein is expressible by the formula $2^m \times 3^n$, where m and n are integers, and that the number of residues of any individual amino acid was expressible by the formula $2^{m'} \times 3^{n'}$ where m' and n' could be integers or zero. Now here was a 'magic formula', which, if true, could lead to all kinds of mathematical speculation as to how proteins might be assembled in the living cell! But, of course, it wasn't true.[19]

It is important to appreciate that such wild ideas were given every full consideration and not discarded out of hand. Bergmann was on the staff of the prestigious Rockefeller Institute in New York and Niemann was at CalTech and well regarded by Linus Pauling, with whom he collaborated on several projects. Even today, Bergmann is considered by some historians to have been a brilliant scientist and his theory a valuable contribution at the time.[20,21]

Dorothy Wrinch and the cyclol theory

The most forgettable of all the fruits of the 1930s' harvest, not really worth more than a footnote, was a theory built on nothing, no training, no relevant skills. But it had its brief moment in the history of protein science, more in the public limelight than most, partly as a result of the sheer bravura (chutzpah) of the author, and partly because of the support of Irving Langmuir, one of the great physical chemists of the early twentieth century and held in some awe by the protein community.

Dorothy Wrinch was an English mathematician, who (among other attachments) was at one time infatuated with Bertrand Russell and his philosophy. She had no formal training in chemistry and at best a rudi-

Fig. 10.2 Dorothy Wrinch giving a talk in Cambridge in 1933. From the biography of Wrinch by P. G. Abir-Am.[22] (Original source: C. H. Waddington; courtesy of Gary Werskey.)

mentary comprehension of the rules of evidence in scientific research. She was arrogant and felt persecuted when criticized, but in retrospect her miseries seem self-inflicted.[22]

Wrinch's specific proposal was a geometrical construct requiring rejection of the simple peptide link between amino acid residues of the protein backbone. It was replaced (by means of a chemical change analogous to lactam–lactim tautomerism in other organic compounds) by a three-pronged arrangement which lends itself to construction of six-membered rings and creation of honeycomb-like polyhedra. A barrage of papers appeared,[23] claiming not only to account for molecular compactness, but also to explain the supposed molecular weight classes suggested by Svedberg. (See ref. 24 for a review by Wrinch lacking the tone of stridency of her shorter papers.)

Following her cyclol proposal, Wrinch began to bombard famous people with letters, seeking interviews and collaboration. Linus Pauling was her target in the summer of 1936, for example, but no meeting between them resulted until 1938.[25] However, she was more successful with Irving Langmuir, who may have been receptive to Wrinch's approach because she suggested new experiments, adaptation of his surface balance measurements to protein layers, firing his imagination with a promise of exciting new vistas for exploration. In any case, Langmuir asked Wrinch to visit him in Schenectady in December 1936 and initial experiments with the surface balance were done and were successful. 'This should have great value as a biological tool: very likely it will find a place in the diagnosis of disease,' he said in a lecture.[26,27]

It is difficult to understand in retrospect how Langmuir could have remained an advocate for long. Wrinch's structures were sterically impossible. She had focused geometrically on the polypeptide backbone, with amino acid side chains represented simply by the letter 'R'. When 'R' was replaced by actual atoms, there turned out to be not enough room for them! Neurath and Bull had demolished the cyclol theory on this basis already in 1938,[28] but no third party criticism should have been needed, for Langmuir had previously always demonstrated exquisite pictorial imagination and a unique sense of how molecules occupy space.[29,30]

No matter how irrational it may seem now, the involvement with Langmuir helped to give Dorothy Wrinch a hearing all around the world—Dorothy Hodgkin, for example, mentions the 1938 Cold Spring Harbor symposium, a particularly prestigious venue, where Wrinch 'captivated everyone with her enthusiasm'.[31] And the Wrinch–Langmuir association did lead to a brief period of interaction of Langmuir with the mainstream of the protein science community, with ultimate benefits that are detailed in Chapter 12.

Protein denaturation

Much more productive ideas came from the study of protein denaturation, a phenomenon known for a long time as a change of state of a protein, inducible by heat, extremes of pH, or addition of a variety of chemical reagents. Textbooks around 1900 defined it as loss of solubility (or 'coagulation') and it was already then understood to involve no change in molecular weight.[32] In more modern times the definition would have included loss of biological activity, loss of ability to crystallize, spectral changes, and change in accessibility of sulphhydryl and other reactive groups to chemical probes.

Another early defining attribute of denaturation would have been its irreversibility. Boiled egg white, for example, the most familiar of all denatured proteins, can be brought into solution by acid or base, but reprecipitates when the pH is brought back to neutrality—the very property that defines an 'albumin' is permanently lost. This kind of loss was thought to be quite general and discouraged research on denatured proteins. In the words of M. L. Anson and A. E. Mirsky, 'it has been taken as almost axiomatic that coagulation is irreversible', and therefore obviously 'of little physiological interest'. But then Anson and Mirsky themselves changed all that. They found in 1925 that denatured haemoglobin can (with care) be restored to where it can again be crystallized and combine with oxygen. Denaturation of serum albumin, trypsin, and subsequently many other proteins was also found to be reversible. Egg albumin is an exception, not the rule.[33,34]

This discovery changed the whole picture. Reversibility implies that the transition is an equilibrium process and that native and denatured structures are thermodynamically determined functions of the 'state of the system'. One could now analyse the effects of temperature, chemical reagents and other factors by rigorous thermodynamic methods, in terms of *shifts in pre-existing equilibria*, and thereby the experimental study of denaturation, by chemical or optical probing of denatured and native states, became an invaluable guide to *thinking* about protein structure.[35] Many subsequently influential papers began with exploration of the native–denatured transition: for example, Mirsky and Pauling's introduction of the hydrogen bond into the protein chemist's vocabulary in 1936 (see the following chapter) came in a paper entitled 'On the structure of native, denatured, and coagulated proteins'.

The earliest interpretive 'thinking' effort came from China in 1931, from the laboratory of Hsien Wu, a remarkable scientist, whose career spanned the turbulent period of China's transitions from Manchu dynasty to republic to communism—with Japanese occupation en route. (For a biography, see Edsall.[36]) Wu's view of denaturation was simple and uniquely lucid: he saw the native protein as an organized compact molecule, 'not a flexible open chain of polypeptide'; he saw denaturation as destroying that organization, creating a more disordered molecule. Wu assumed that the native organization was due to non-covalent forces of attraction, mainly between polar groups. He included ionic groups within the 'polar' category, but also saw the need for involvement of peptide groups to create an ordered polypeptide backbone.[37] Hydrogen bonds were not yet widely known. Had they been, Wu would undoubtedly have considered them a likely possibility. Hydrogen bonds were in fact invoked in 1936 by Mirsky and Pauling as the principal non-covalent force of attraction, in what would prove to be the most influential work on denaturation of the decade (see Chapter 12). Mirsky and Pauling were apparently not aware of Wu's earlier work; at least they make no reference to it. Pauling has called his 1936 paper 'the first modern theory of native and denatured proteins'.[38]

As it turned out, the hydrogen bond theory was not the last word. Denaturation continued for a long time to be a lively topic, both for research and for thinking about protein structures. A generation after Mirsky and Pauling, it led to another historical milestone, a scholarly review by Walter Kauzmann of all the forces that might be involved in maintaining native protein structures.[39,40] He came to the unexpected conclusion that the most important energetic driving force was provided by 'hydrophobic bonds' and thereby he opened up an entirely new factor in the understanding of how proteins work, which will form the subject of Chapter 12. Current wisdom would probably say that both hydrogen bonds and the hydrophobic force are involved in maintenance of the native structure of proteins, though in rather different ways.[41] (See footnote in Chapter 12, p. 140.)

Polar and non-polar groups

In Chapter 5 we focused on the gradual realization of the existence of ionic groups in protein molecules—whole electrostatic charges fixed to particular atoms. In this context we spoke of molecular *dipoles*, molecules that were neutral *in toto*, but contained two full ionic charges, one positive and one negative, both fixed to particular groups—for example, $-NH_3^+$ and $-COO^-$. We also pointed out that the term 'dipole' when used with reference to proteins was really not appropriate and that 'multipole' would be more realistic, because protein molecules could contain huge numbers of ionic charges per molecule, which would exactly neutralize one another only at a single pH, called the isoelectric point.

We are now, in this section of the book, further along in the history of chemistry as a whole; electrical polarity has extended to cover a broader territory. In particular, every chemical bond is known to have electrical polarity; nearly every molecule is a 'dipole' in the sense that its bonds are not arranged with perfect symmetry. Within a protein molecule, C–C and C–H bonds have minimal asymmetry and thus hydrocarbon moieties have little polarity and are usually designated as 'non-polar', whereas oxygen and nitrogen atoms are intrinsically electronegative and C–N and C–O bonds are therefore 'polar'. (Among whole molecules, H_2O is notable for its relatively strong polarity, positive at the location of the hydrogen nuclei and negative at the oxygen end—though not nearly as polar as it would be if there were whole charges on the H and O atoms.) These polarities provide the basis for the secondary forces that loom so large in the chemistry of proteins and the solvent H_2O in which they are normally immersed.

About ionic bonds

Thinking about the folding of polypeptide chains into compact structures, whether it was done on the basis of experiment (that is, denaturation data), or purely speculatively, inevitably required an understanding of secondary forces of intramolecular attraction that would be weaker than primary valence bonds—intactness of primary bonds is a given fact.

An old idea, which used to be the only plausible possibility, was the idea of attraction between positive and negative charges ('ionic bonds'). There were plenty of charges along the chain. Would they tend to arrange themselves to satisfy such attractions? This idea was still considered valid in the 'visionary' period we have been portraying in this chapter, despite mounting evidence against it, which suggested that in globular proteins the ionic charges were mostly on the molecular surface, making ion-dipole bonds to water molecules. Hsien Wu,[37] for example, had in 1931 given ionic bonds as much weight as secondary bonds between peptide groups. And in discussions of the fibrous proteins, links between glutamyl and lysyl side chains were widely thought to be needed for transverse bonding between adjacent polypeptide chains.[42] Bernal in 1939 unambiguously dismissed ionic bonds as 'plainly out of the question' because they would 'certainly hydrate', but he was unusually perceptive.[43] Electrostatic forces continued to be invoked sporadically for some time after that and the concept of them as a conceivably important energetic factor was not truly laid to rest until 1949.[44]

Since then the twentieth century concepts of hydrogen bonds and hydrophobic 'bonds' have proved to be the most productive tools of this kind and we have given them detailed treatment in the following individual chapters.

Overall perspective

To put the concepts of secondary bonds into proper perspective, we should note that they came into universal use (and continue to be used today) not only in thinking about protein structure itself, but also in interpreting how protein structure is used to form binding sites for ligands, active sites for enzymes, and the like.

To put reversible protein denaturation as a whole into proper perspective, we need to look forward to the 1960s and beyond, after the molecular mechanism for protein synthesis was understood and the relation between the gene and amino acid sequence was elucidated. How does nature proceed from the linear sequence to the native three-dimensional structure? The reversibility of denaturation suggested that it required no new principles; folding of a nascent polypeptide chain is analogous to *refolding* of a denatured protein. 'Sequence determines structure' became a dogma of protein science.[45]

CHAPTER 11

Hydrogen bonds and the α-helix

Such combination need not be limited to the formation of double or triple molecules. Indeed the liquid may be made up of large aggregates of molecules, continually breaking up and reforming under the influence of thermal agitation.

W. Latimer and W. Rodebush, 1920,

in reference to liquid water[1]

The covalent bond, formed by electrons shared between atoms, was a concept created in 1916 by G. N. Lewis at the University of California[2,3] and was therefore still quite new in 1920, when Wendell Latimer and Worth Rodebush, also at the University of California, first proposed hydrogen bonds, defining them as secondary interatomic links, weaker than valence bonds.[1] From the very start they recognized the electrostatic origin of the bonds, noting that they form between atoms of high electron affinity, such as oxygen and nitrogen. Likewise, they recognized the special importance of hydrogen bonds in liquid water. The core of many ideas that were later developed quantitatively by others can be found in embryo in this first paper on the subject.

The structure of water

The concept of hydrogen bonds was famously exploited in 1933 by J. D. Bernal and Ralph Fowler to create their model of the structure of liquid water, a milestone in the *history* of physical chemistry that proved to have a huge indirect impact on protein science.*[4]

---

* The authors avoided the use of the word 'bond' in relation to the phenomenon. The idea of shared electrons 'bonding' atoms together was still fairly new and there would have been reluctance (especially by physicists) to extend the meaning of

The origin of the work is interesting:[6] it arose from the presence of dense fog at Moscow airport after a theoretically oriented conference, in which Max Born, Walter Heitler, Fritz London, and other notable figures were participants. Bernal and Fowler were scheduled to depart at 4 a.m., but the fog prevented the plane from taking off until 4 p.m. The airport was new and there was not even a place to sit down. Bernal and Fowler walked up and down, 'prevented from doing anything but thinking and talking'. Naturally enough, their talk turned to the cause of their discomfort—why did water molecules form this opaque mass of droplets known as 'fog'? Fowler (a theoretical thermodynamicist of considerable stature) knew nothing about water molecules and Bernal had to explain the subject to him. As the hours went by, the discussion turned to other aspects: the quartz-like structure of ice, the densities of ice and water, etc. It is one way (a rather attractive way) that history can be made—by accident, not by purposeful determination.

Bernal and Fowler's theory views the H_2O molecule as essentially the same in liquid water (and in fog) as in steam, with an O–H distance of $0.96\,\text{Å}$. Experimentally, on the basis of an X-ray scattering curve for liquid water, they deduced that the O–H bonds are directed towards oxygen atoms of adjacent H_2O molecules, with an O–O distance of $2.76\,\text{Å}$. The hydrogen bond is seen not as an equal sharing of a hydrogen atom between two oxygen nuclei (nowhere near equidistant), not analogous to the sharing of electrons in a covalent bond. It is a polar bond, the H atom having positive polarity and being strongly attracted to the negative polarity of the unbonded electron pair of the distal water molecule. An elaboration came a little later from Bernal and his student, Helen Megaw. Here is introduced the notion that hydrogen atoms in bulk liquid water can jump in concerted fashion between the two oxygen atoms that they link together—this is how fluidity can be maintained without much disruption of the overall tetrahedral (quartz-like) arrangement of the water molecules.[7]

These are pioneering papers, the theory encompassing not only the 'structure' of liquid water, but many related topics. They provide a general

the term to a weaker form of attraction. Bernal and Fowler also make no reference to Latimer and Rodebush's earlier paper—they may not have been aware of it. According to J. Donohue,[5] the 'reality' of hydrogen bonds did not penetrate to a broad chemical audience until after the publication in 1939 of Pauling's book on the nature of the chemical bond[10] (see p. 124).

relationship between hydrogen bond energy and O–O distance in molecules other than H_2O; they also discuss hydration of dissolved ions in general terms. In relation to this, they point out that H^+ must exist in water as the hydrated ion H_3O^+. With this in mind, they account quantitatively for the anomalously high ionic mobility of H^+ and OH^- in water (five times higher than expected), by virtue of what is now generally known as the Grotthus mechanism.[8]

Hydrogen bonds and protein folding

Meanwhile Linus Pauling had begun graduate work at the California Institute of Technology (1922–1925) and developed the interests that would lay the foundations for his career, the experimental determination of the structure of crystals by X-ray diffraction and its theoretical interpretation. He quickly absorbed quantum mechanics and other new theoretical concepts and applied them all to the problems of atoms and the bonds between them. He was impatient for success; his published work often went beyond the immediate goals of his projects. For example, his first paper on the shared electron bond already spills over into hydrogen bonds and recognizes their electrostatic origins

A longer definitive paper[9] followed soon after and Pauling eventually became the acknowledged authority on all chemical bonding with his book, *The nature of the chemical bond*, the first edition of which was published in 1939. As expected, the book contains a chapter on hydrogen bonds, but there is as yet no mention of proteins. Pauling's early career was tilted strongly towards the inorganic side of chemistry.

Pauling's interest in proteins, according to his own recollections,[11] stemmed partly from financial considerations—support from the Rockefeller Foundation that would not have been given for his work on the crystallography of inorganic compounds—and partly from stimulus provided by the review of Anson and Mirsky on the reversibility of the denaturation of haemoglobin and other proteins.[12] Alfred Mirsky was at the Rockefeller Institute at the time and, when Pauling felt the need for a co-worker with expertise in the handling of proteins, he was a logical choice. The Rockefellers even agreed to pay his salary.

The outcome was a comprehensive paper on protein denaturation. Hydrogen bonds had by now become part of the fashionable dogma— even the unprepared mind could appreciate the anomalous properties of liquid water and how hydrogen bonds were able to explain them. It was thus natural for Pauling to invoke hydrogen bonds to provide the cohesive force needed for the organization of native proteins. In the provocative style that would become characteristic of the man, he would not admit to any other plausible possibility.[13]

This belief in the inevitability of hydrogen bonds as structural determinants was not universal. For example, a scholarly review by Hans Neurath and several distinguished co-authors cautiously characterizes the nature of the cohesive bonds as still unknown as late as 1944.[14] But Mirsky and Pauling, in their paper, indicate no need for such caution. After defining what denaturation is and explaining its reversibility, they wade right in without preliminaries: 'The polypeptide chain is folded into a uniquely defined configuration, in which it is held by hydrogen bonds.' Two kinds are mentioned: those between nitrogen and oxygen atoms of the polypeptide backbone and those between side-chain amino and carboxyl groups. It is recognized that not all of the side-chain groups will be used in forming bonds within the molecule; some will be free on the surface of the molecule.

The paper included the known geometric characteristics of hydrogen bonds in general, e.g. an N–H–O distance of about 2.8 Å, shorter than it would be for unbonded N and O atoms. The energy of hydrogen bonds is given as around 5–8 kcal/mol, the lower value being identified as approximately correct for an N–H–O bond 'in proteins'. Side-chain hydrogen bonds are assigned a similar energy. The typical sort of activation energy that is observed experimentally for denaturation, 150 kcal/mol, is thus accounted for by the need to break about 30 side-chain hydrogen bonds.

With hindsight we can see that this thermodynamic analysis was faulty. Hydrogen bonds within protein molecules are not actually 'broken' in the typical denaturation experiment in aqueous solution, but replaced by hydrogen bonds to water molecules—or by bonds to 'denaturant molecules', when a chemical denaturant is used. Actually, Mirsky and Pauling imply this when they explain the probable mechanism of the action of denaturants: urea and many other known denaturants 'form hydrogen bonds with the protein side chains, which are thus prevented from combining with each other and holding the protein in its native configuration'.[13] The logical conclusion is that the *net energy change* should then be about zero, but the text is ambiguous in this context and considerable confusion about the strength of hydrogen bonds remained in the literature for many years. The internal hydrogen bond theory *per se* remained the dominant theory for the creation of collapsed protein molecules for at least two decades.

The α-helix and the β-sheet

After the Mirsky–Pauling paper on protein denaturation, Pauling began to try to devise a more precise definition of the bonds that might be essential to the structure of a native protein. He started with Astbury's speculative α and β structures for the polypeptide chain backbone,[15] which he was able to dismiss out of hand, but he was unable to come up with anything more convincing.

At this point Robert P. Corey, already trained as an X-ray crystallographer, joined Pauling's group—like Mirsky, he came from the Rockefeller

Fig. 11.1 Linus Pauling lecturing, with molecular models at his side. (From the Ava Helen and Linus Pauling papers, Oregon State University special collection.)

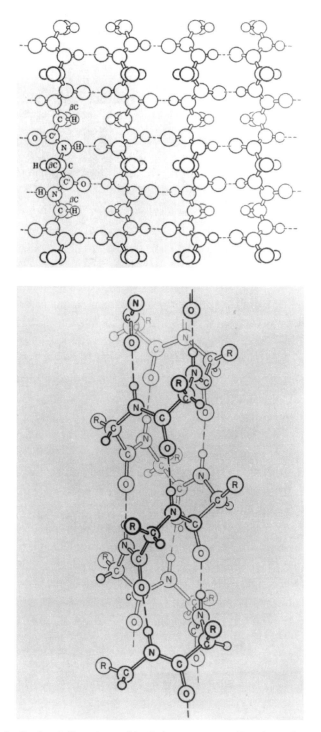

Fig. 11.2 Pauling's α-helix and one of the β-sheet structures. (Based on refs. 16 and 17; courtesy of L. Pauling and R. B. Corey.)

Institute. The two of them agreed to embark on a unique project, to solve the problem of bond definition not by analysis of proteins themselves, but by determining the structures of crystalline amino acids and simple peptides—that is, compounds of relatively simple structure, containing what may be called fragments of proteins, often including intermolecular hydrogen bonds. These structures could be determined rigorously by X-ray analysis and yielded atomic distances and bond directions that could be used in the building of models for the much more complicated structure of a protein. With the addition of general theoretical principles that were second nature for someone with Pauling's background, they established the rules that a folded polypeptide would need to obey.

The now famous α-helix and β-sheet structures[16,17] were the result of this work, which, with interruptions by the war, took more than a decade to complete. One of the key conclusions was that each peptide group, together with the two proximal carbon atoms it joined, must be in a planar configuration. A result that applied specifically to the α-helix was that no reason emerged to assume an integral basis for the helix repeat, so that this intuitively attractive limitation could be abandoned—their final conclusion was that a structure with 3.7 amino acid residues per turn of the helix would be optimal.[16*]

It was a remarkable piece of work. No experimental data from protein fibres or crystals were used to arrive at the result. No startling new principles were invoked; the result was based entirely on the confident prediction that the structural elements would be the same for proteins and for the model compounds.

We have emphasized that no experimental protein data were used to arrive at the result. In fact, to make the history complete, it is necessary to point out that the predicted α-helix *contradicted* one of the most striking features of X-ray data from protein fibres such as keratin and muscle—a strong and ubiquitous reflection at 5.1-Å separation, where the theoretical helix called instead for a 5.4-Å reflection. This discrepancy might have

---

* Dr H. Branson, a visiting professor, checked Pauling's calculations and found that there were actually two helical structures consistent with the stringent Pauling–Corey conditions, the second being flatter than the α-helix, with 5.1 residues per turn. In the case of the β-structure, intended to apply to the laterally hydrogen-bonded, fully extended, polypeptide chains of silk fibroin or stretched keratin, Pauling and Corey identified two plausible 'pleated sheet' configurations, one with parallel and one with antiparallel alignment of the chains.

been expected to lead to automatic rejection of the theoretical structure. Did not the most sacred rule of theoretical research demand perfect agreement with experiment? In fact, acceptance quickly became all but universal, for which three reasons may be cited.

1. The theoretical line of reasoning and its consistency with the model compound data were so powerful that they could not be denied. And by this stage in the history of protein science the yearning for special imaginative ideas, exclusively valid for proteins, had disappeared.

2. On-going studies at the industrial firm of Courtaulds in England indicated that fibres made from synthetic polymers of L-amino acids, poly-γ-methylglutamate and poly-alanine, had a helical structure and their X-ray pictures, unlike those from protein fibres, agreed almost exactly with the predictions of the Pauling model, with a layer line at 5.4 Å and none at 5.1 Å.[18] This work was not published until a couple of years later, but news of it was circulating among protein crystallographers about the time of Pauling's publication.

3. Max Perutz at Cambridge, one of the first to recognize the 5.1 Å/5.4 Å discrepancy, supported the Pauling structure on the basis of new evidence obtained in his own laboratory, namely the discovery of a spacing of 1.5 Å along the longitudinal axis, which could not be interpreted other than as a measure of the repeat distance of amino acids in that direction[19] The measured value corresponds to 3.7 residues per turn of the helix and is 'in perfect accord' with the Pauling model for the α-helix and incompatible with any other regular (that is, helical) structure that had been proposed. Perutz lists five, including structures he himself had proposed just a year or two earlier.[20] The 1.5-Å distance was observed in haemoglobin crystals, muscle fibres, and another synthetic polypeptide obtained from Courtauld, which was studied as an oriented film instead of as a fibre.[21]

The 5.1 Å/5.4 Å discrepancy remained of course, but now seemed to suggest some peculiarity in the keratin class of proteins, rather than a fault in the molecular model. Max Perutz suggested the problem to his colleague, Francis Crick, who went on to show that the intrinsic undistorted α-helix can be appreciably deformed, to improve the packing of amino acid side-chain 'knobs' when adjacent helices associate with each other—leading to what has become known as a 'coiled coil'.[22,23]

Thomas Hager, Pauling's biographer, has called Pauling's structural work 'one of the most extraordinary sets of papers in the history of twentieth-century science'; he states that the difference between the observed 5.1-Å reflection and Pauling's predicted 5.4-Å reflection 'was the only thing standing between him and the honour of being the first man to describe the structure of a protein',[24,25] which is obviously rubbish. Pauling had not described the structure of any protein, nor did he claim to have done so. Nevertheless, it is true that the α-helix somehow captured the public imagination and became an emblem, an icon for all of molecular biology and its inherent promise. Even today, fifty years later, much of the general public may still think of the 'α-helix' and the DNA 'double helix' as basically the same thing—the molecular spiral as an essential element of the science of life.

CHAPTER 12

Irving Langmuir and the hydrophobic factor[1]

Because of the large number of peptide groups and hydrophobic groups in nearly all proteins, it is likely that hydrogen bonds between peptide links and hydrophobic bonds are by far the most important in determining the over-all configuration of the protein molecule.

W. Kauzmann, 1959[3]

Walter Kauzmann, in two reviews in 1954 and 1959, made protein chemists aware of the hydrophobic force and suggested that it was energetically more important than hydrogen bonds in folding proteins in aqueous solution into the compact shapes that most of them adopt.[2,3] The same point had actually been made nearly twenty years earlier: how it happened and then came to be virtually forgotten[4] makes a good story and provides a picture of an interesting period in protein history, more confrontational than anything we have seen since.

Definition and early history

Hydrophobic antipathy is another manifestation of hydrogen bonds, but the important hydrogen bonds in this case are those between water molecules in the solvent and not bonds that directly link protein groups internally to one another. Apolar surfaces, normally hydrocarbon entities in the case of proteins, are forced into close proximity (in effect squeezed out) by the attraction of water molecules for one another, which dominates the scene energetically—no 'like to like' *attraction* between apolar molecules is importantly involved in the overall energetics. The effect becomes especially interesting in the case of proteins (or any other substance in the so-called *amphiphilic* category) where both hydrophilic and hydrophobic chemical groups may be present on the same molecule,

131

one seeking contact with water, the other avoiding it—a dichotomy leading quite naturally to preferred molecular orientations or (within a single molecule) to unexpected conformational contortions.

The first use of the hydrophobic concept as a factor in molecular orientation was made over a hundred years ago, in relation to surface tension, by the German physical chemist Isidor Traube,[5] who showed that many organic solutes are adsorbed at a water/air interface, with the polar ends of molecules in the water and non-polar parts sticking out into the air.[6] Irving Langmuir subsequently became the true master analyst of events at such an interface: his 1917 paper on surface layers of organic molecules is a *tour de force* in which Langmuir displayed an extraordinary pictorial molecular imagination, by means of which he was able to interpret measurements with a simple surface balance, so as to visualize in his mind exactly what individual molecules in an adsorbed film were doing. He recognized the permanent 'kink' caused by a double bond in an aliphatic hydrocarbon chain, to give just one example.[7] In 1932 Langmuir was awarded the Nobel Prize for chemistry for this work. None of his conclusions has ever been seriously challenged.

Another early field of application was to soap micelles in aqueous solution, where we are dealing with multimolecular aggregation of amphiphilic monomers. The 1936 treatise of G. S. Hartley on this subject was particularly articulate in stressing the distinction between the hydrophobic mechanism for aggregation and 'like to like' attraction.[8] Along the same lines, the most obvious biochemical application of the hydrophobic idea is to biological lipids and cell membranes. A famous paper by Gorter and Grendel (1925) on this subject was explicitly derived from Langmuir's 1917 paper and presented compelling evidence to show that phospholipid molecules exist in nature as bilayers (which can be thought of as extended micelles), with the hydrocarbon parts facing each other in the middle.[9,10]

These seminal works on micelles or bilayers had little or no influence on the exploration of theoretical causes for the folding of proteins, for the simple reason that extension of the hydrophobic idea to proteins was not obvious. In micelle or bilayer formation one is dealing with solute molecules that are homogeneous or nearly so, with very long aliphatic hydrocarbon chains, which one can easily visualize, for example, as lining up in parallel. On the other hand, the non-polar moieties of protein amino acid side chains are not only short in length, but some are aliphatic and some aromatic. Furthermore, they are all mixed up with polar groups along the polypeptide chain, rather than neatly segregated.

Weeks of fierce debate: J. D. Bernal and others challenge the cyclol theory

The original application of the hydrophobic principle to proteins was made in 1938 by none other than Irving Langmuir, who was, of course, a chemist of great distinction, but at the time a complete outsider to the protein community. How did he become involved? Strangely, it arose by the pathway of Dorothy Wrinch's despised cyclol theory, which Langmuir for a time advocated (see Chapter 10).

Trying to find rational support for the theory, Langmuir reached back into his past experience, to the work at the air/water interface that we have already mentioned. He recognized that a substantial fraction of any protein's side chains are hydrophobic and that the need to remove these entities from contact with solvent must be a driving force for the folding of polypeptide chains into the compact 'globules' that the soluble protein molecules are known to be. In one of the papers Langmuir goes so far as to make a quantitative estimate: stabilization of 2 kcal/mol will result when two CH_2 groups move next to each other away from contact with water. All this was said in the context of the cyclol hypothesis. Given the thermodynamic need to 'collapse' into a small space, the polypeptide backbone must find some way to make it sterically possible, some structural device that might otherwise not seem natural.

Langmuir had first spelled out the hydrophobic theory for protein folding at an exceptionally lively Cold Spring Harbor symposium on protein chemistry in the summer of 1938,[12] but his paper was mostly devoted to protein monolayers adsorbed to a surface and there is no indication that Langmuir struck a responsive cord among the assembled protein chemists on that occasion. Later that year, however, he came to England, now as a vigorous champion of Wrinch's cyclol hypothesis, and thereby he triggered a fierce debate, in which J. D. Bernal emerged as the intellectual leader. While almost everyone in the protein community (at least in England) was beginning to be made angry by the cyclol hypothesis and the uncritical manner in which it was touted, Bernal felt more personally challenged than anyone else, for the problem so casually and often arrogantly addressed by Wrinch (the intimate arrangement of atoms

Fig. 12.1 Irving Langmuir hiking in the Adirondack Mountains: Langmuir was intense and ambitious about everything he put his mind to. (Photograph kindly provided by Dr George Gaines of the General Electric Co.)

within protein molecules) was the problem to which Bernal was hoping to devote a lifetime of research. He was particularly angered by Wrinch's claim[13] that preliminary X-ray data for insulin, published by Bernal's Ph.D. student Dorothy Crowfoot,[14] supported the cyclol structure—the analysis that led to this assertion was at best sloppy and incompetent, at worst dishonest.[15]

Fig. 12.2 Langmuir, Bernal, and Hodgkin at the 1937 meeting of the British Asociation. (Source: Hodgkin's biography of J. D. Bernal, 1980.[3])

The war of words (all of which took place in London within a few weeks) began at a meeting of the Royal Society on protein molecules on 17 November 1938, where T. Svedberg was principal speaker.[16] Dorothy Wrinch crusaded on behalf of her structural theory, which was severely criticized on the spot by the distinguished biochemist A. Neuberger.[17] Bernal gave a talk about X-ray crystallography at this meeting, but he was not yet involved in the cyclol controversy. He became involved two weeks later, on 1 December, on the occasion of his inaugural address as professor of physics at Birkbeck College. It was a general lecture entitled 'The structure of solids as a link between physics and chemistry', but there is anecdotal information[18] to indicate that his mind was very much on proteins. It appears that the lecture was hastily prepared, slides for it being gathered at the last minute. Bernal went to the Royal Institution to borrow some extra slides but he was collared there by Langmuir (preparing for his own lecture the following week?) who attempted to convince him of the merits of the cyclol theory. Bernal was incredulous at his naiveté and presumably tried to argue with him, for he was detained for hours and almost arrived late for his Birkbeck talk.

A week after Bernal's inauguration came two lectures by Irving

Langmuir, the Pilgrim Trust Lecture at the Royal Society on 8 December and a lecture at the Royal Institution on 9 December—both were published without delay.[19,20] Langmuir came out explicitly with the hydrophobic/hydrophilic principle as a basis for globular protein structure, noted the parallel between globular proteins and soap micelles, etc.—it is a scholarly presentation, though Langmuir only brought it up in the course of his advocacy of the cyclol theory.

During the same period there was a deluge of letters and articles on the cyclol hypothesis in the weekly magazine *Nature*. One issue alone (14 January 1939) had a paper by Langmuir and Wrinch[21] which reasserts that the cyclol hypothesis is 'confirmed' by the insulin X-ray data, and three letters attacking that assertion, the most scathing by Bernal,[22] but another almost as strong by Nobel Prize-winning X-ray crystallographer Lawrence Bragg.[23] Letters to *Nature* were at that time published within a week or two of receipt and were thus a forum for extremely rapid dialogue; that is, these letters were all written after the foregoing lectures. The series of communications to *Nature* contains only a single voice of support for Wrinch, a letter by E. H. Neville of the University of Reading,[24] which ends with the familiar call often heard in support of ideas that lack a foundation: let anyone who wants to argue with Wrinch propose a better structure than hers.*

Soon after came a crucial lecture by Bernal at the Royal Institution (27 January 1939), entitled 'Structure of proteins'.[26] In it Bernal, putting aside the cyclol debate, had surprisingly become an advocate himself of the hydrophobic principle. In Bernal's own words: 'Langmuir has used this picture as a justification of the cyclol cage hypothesis, but it is strictly quite independent of it.'

It is a fascinating phenomenon: in most people's minds an angry rejection of the cyclol hypothesis would probably lead to automatic rejection of everything associated with it, but Bernal managed to find the gem of truth within the dross. He has thought about and been convinced by the hydrophobic mechanism, which had been advocated by Langmuir in the very same paper that Bernal castigated two weeks earlier. The words Bernal

---

* It turns out[25] that Neville was Wrinch's 'intimate friend' and 'adoring lover', so his status as an objective outsider is questionable. Later, in the 1950s, Neville proposed marriage to Wrinch, but her ardour had evaporated. It would have been her third marriage.

now uses are his own, not mere reiterations of Langmuir's.[27] He explicitly makes a point that Langmuir does not: 'Ionic bonds are plainly out of the question, as they would certainly hydrate', putting down the likelihood of polar links as a mechanism for folding. Other direct quotations from Bernal's lecture are: 'The behaviour of the hydrophobe groups of the protein must be such as to hold it together.' 'In this way a force of association is provided which is not so much that of attraction between hydrophobe groups, which is always weak, but that of repulsion of the groups out of the water medium.' Bernal has obviously penetrated to the essence of the matter.[28]

In the same paper,[26] Bernal of course also discussed X-ray diffraction and the structural information derived from it, and ultracentrifuge studies and the like as well. This is a pioneering and prophetic paper, a glimpse into a rare moment when one man's insight was able to encompass simultaneously all the strands of a complex problem, much of which the rest of the protein community would not understand for another twenty years. The paper was reprinted in *Nature* in April 1939 with only minor changes and was thereby made available to the world at large, and many readers (or listeners at the lecture) must have absorbed and stored away the gist of it. We have been able to find no further explicit endorsements in print, but it is plausible that the idea was henceforth 'in the air', as Kauzmann claimed, mentioned in lectures or informal discussion.

Intervention of the war

The year 1939 saw the beginning of the war in Europe and the effective end of this heated debate about the underlying cause of protein structural compactness. Bernal became a scientific advisor to the British government, as we have already noted. Langmuir was recruited to work on smoke filters for gas masks and the like. General dissemination of the hydrophobic idea ceased. We can find no mention of the idea after the war until Kauzmann's reviews.

Even Bernal himself seems to have forgotten about his pre-war enthusiasm for the hydrophobic principle. As late as 1958 he gave the introductory lecture at a general discussion of the Faraday Society.[29] In the course of it he states that 'It is hardly worth recalling here the kinds of forces we shall have to deal with' but then goes ahead and recalls them anyway: in order of strength they are (1) covalent, (2) ionic, and (3) hydrogen bonds, where C=O...H–N bonds are emphasized. He refers to the importance of hydrogen bonds for liquid water (as he should, being one of the original exponents), but there is no mention of his earlier lucid explanation of why this should be a vital factor in protein folding.

Entropy and enthalpy

The next important contribution to the subject is in itself rather improbable as a potential influence on protein science and it came from an unlikely source, Henry Frank, a Christian missionary in China with a Ph.D. in physical chemistry, who taught chemistry at Lignan University. He was interned by the Japanese after Pearl Harbor, but was repatriated in 1942 as part of an exchange, and was eagerly hired as an instructor at the University of California, to do some of the teaching left uncovered by permanent faculty who had gone to do war work. He was, of course, also free to do research and returned to a subject that had long been on his mind, the derivation by statistical mechanics of equations for thermodynamic data of liquids. The idea was to use a popular (though approximate) approach, in which entropy values are interpreted in terms of 'free volume—that is, the volume over and above the volume occupied by atoms themselves. This was an attractive concept for Frank because it permitted (in his own words) a 'pictorial interpretation'.[30] The first paper dealt with monatomic crystals, the second with pure liquids, the third, with computing assistant Marjorie Evans as co-author, focused on liquid mixtures.[31] Here aqueous solutions immediately stand out as anomalous, for ionic solutes as well as for the non-polar ones which are the main concern here. For non-polar solutes the most striking common feature is a *negative entropy of mixing* and an absence of the positive heat of mixing that should accompany the breaking of H_2O–H_2O hydrogen bonds. These thermodynamic characteristics indicate that hydrogen bonds disrupted by insertion of the non-polar solute do not remain broken, but instead become rearranged in a more restricted, more ordered pattern. Berkeley's most famous chemist, G. N. Lewis, with whom Frank discussed the work, proposed the name 'iceberg' for these ordered regions, with no implication of any detailed resemblance to the structure of an ice crystal, but simply as an illustration of the kind of organization that the thermodynamic data required.

Henry Frank knew nothing about protein chemistry, nor probably about orientation of amphiphilic solutes at surfaces—he does not even use the word 'hydrophobic' when referring to 'rare gases and other non-polar molecules'. Yet his work had a catalytic effect. It was in a sense an extension of the definition of the word 'hydrophobic', an illuminating clarification. This 'pictorial interpretation', crude as it was, captured the imagination of protein chemists—it provided evidence for the dispro-portionate strength of hydrogen bonds in water and engendered con-fidence in the hydrophobic concept. Max Perutz, for example, who had been an early research student of J. D. Bernal and could in principle have learned about hydophobicity from him, has the following recollection (Perutz, personal communication):

> I cannot remember what Bernal said about hydrophobic groups in
> proteins but I do remember being very impressed by Kauzmann's review
> because he gave a reason why hydrophobic groups would be buried and
> drew my attention to the crucial paper by Frank and Evans, which
> otherwise I would never have seen.

Concluding comment

It is easy to underestimate the importance of the hydrophobic principle and to think that all that counts is the structure itself and that the ques-tion of whether one force or another dominates in forming the structure is a mere technical quibble, of the kind enjoyed by theoretical physical chemists. On the contrary, the matter is at the very heart of protein science. This can be illustrated by reference to a subject that is increasingly attract-ing attention, the problem of trying to predict three-dimensional struc-ture on the basis of the amino acid sequence of a protein, plus general rules that govern polypeptide folding and internal organization.

Harold Scheraga, professor at Cornell University, was one of the first to try his hand at structure prediction, but he was also a great enthusiast for the hydrogen bond theory of protein folding, one of the last to favour it. In 1960 he predicted a hypothetical three-dimensional structure for the protein ribonuclease, based on the primary sequence of its 124 amino acid residues, which had just recently been determined. The predicted structure, depicted in detail with atomic models, had the protein held together by internal hydrogen bonds between side-chain polar groups.[32] This of course put many polar side chains in the centre of the structure, and left most of the hydrophobic groups dangling out at the surface. As it turned out, 1960 was also the year when the first high-resolution three-dimensional structure obtained by X-ray crystallography was published. In this, as in all the other protein structures published since then (including ribonuclease), the reverse situation prevails: charged groups and most polar side chains are at the surface, in contact with water, and most hydrophobic groups are 'inside'.

Could anything be more convincing? Scheraga made no stupid errors (of the sort made by Dorothy Wrinch); he constructed his helices exactly as Pauling prescribed, etc. He did everything right in his model building except in the assumption he made at the very start, as to the nature of the bonds and forces that dominate in making a protein what it is. The same principle would apply to theoretical analyses of functional properties (for example, substrate binding, transport in and out of the cell)—they would confuse and mislead if based on wrong assumptions. All interactions of proteins are governed by the same forces and cannot be understood even approximately without appreciating the thermodynamic dominance of hydrophobicity.

It is even fair to go beyond proteins. The hydrophobic force is the *energetically* dominant force for containment, adhesion, etc., in all life processes. This means that the *entire* nature of life as we know it is a slave to the hydrogen-bonded structure of liquid water. This now commonplace conclusion was not generally understood until after 1960. Historically, the role of the hydrophobic 'bond' in protein structures was the trigger for the wider comprehension that followed.*

---

* This universality of the hydrophobic effect makes it worthwhile to give a more refined account of the energetic principle, with more attention to the subtle interplay between different kinds of hydrogen bonds. We have done this in note 33 on p. 275.

CHAPTER 13

Three-dimensional structure

The crystals were hexagonal bipyramids, 2 mm long or more, prepared by John Philpot while he was working for a short time at Uppsala. He had left his preparations in the refrigerator while he was off on a skiing holiday, and on his return was astonished to find how large his crystals had grown. He showed them to Glen Millikan, a visiting physiologist from California and Cambridge, who said 'I know a man in Cambridge who would give his eyes for those crystals'.

That night, Bernal, full of excitement, wandered about the streets of Cambridge, thinking of the future and of how much it might be possible to know about the structure of proteins if the photographs he had just taken could be interpreted in very detail.

> Dorothy Hodgkin, 1968, reminiscing about
> the first crystals of pepsin[1]

Single crystals: the route to atomic resolution

Bernal and Crowfoot's report in 1934 on X-ray photographs of pepsin is a classic, a historical gem that illustrates the excitement that was generated by the mere *potential* of getting structural information at high resolution, directly from the optical interferences and enhancements within a single crystal:[2]

> Now that a crystalline protein has been made to give X-ray photographs, it is clear that we have the means of . . . arriving at far more detailed conclusions about protein structure than previous physical or chemical methods have been able to give.

A few years later, Bernal's enthusiasm remained undiminished. And results justified the enthusiasm—for example, quoting Bernal in 1939: 'From the beginning . . . the pictures yielded by protein crystals were of exceptional perfection.' Reflections were found at high angles, corresponding to small separations, as small as 2 Å, a distance close to the length of interatomic bonds (for example, 1.54 Å is the typical length of a C–C bond). Protein science had reached what is called 'atomic resolution', where in principle one can distinguish the position of each atom from its neighbours.[3]

The snag was that practical means for using the information locked within the X-ray diffraction pattern were not yet in sight. Bragg's law could only yield 'separations', distances between planes of high electron density in a crystal. In 1935, A. L. Patterson (1902–1966), a New Zealander with a Ph.D. from McGill University in Canada, took a small but important technical step forward by showing that one could obtain a two-dimensional map of actual interatomic distances*—more tangible parameters than distances between planes in a crystal, but still separations rather than absolute coordinates. Patterson himself said it would be 'madness' to think of solving protein structures using this approach.

Thus, although more X-ray pictures of proteins gradually began to emerge, they could give little information beyond defining the 'unit cell' of the crystal, which, at the molecular level, provided no more than a measure of overall shape and size.[5,6] Further progress required not only solution of the inherent mathematical/theoretical problem, but also availability of large single crystals suitable for analysis—as is illustrated by the first words of a serious effort by Bernal et al. to extend the available data:[7] 'We have recently been fortunate in obtaining well developed crystals of chymotrypsin and haemoglobin.' Perutz's subsequent choice of haemoglobin for further study was not initially dictated by haemoglobin's unique physiological or chemical properties, but by the fact that the chymotrypsin crystals he had obtained were twinned and the X-ray data would therefore be more difficult to analyse.

To most people engagement in this kind of research would have posed an unacceptable risk from the point of view of career advancement. There were many who believed well into the 1950s that it could not be done at

---

* To be precise, they were projections of these distances onto a reference plane, still lacking any sense of direction in three-dimensional space. Patterson had received part of his early training at the Kaiser Wilhelm Institute for fibre research in Dahlem and thus represents a rare individual who bridged the gap between the kind of X-ray analysis we discussed in Chapter 6 and the much greater sophistication of single crystal work.[4]

all! For example, in 1954, Professor G. Hägg, in the presentation speech for Linus Pauling's Nobel Prize in that year, can be quoted:

> To make a direct determination of the structure of a protein by X-ray methods is out of the question for the present, owing to the enormous number of atoms in the molecule. A molecule of the coloured blood constituent hemoglobin, which is a protein, contains for example more than 8,000 atoms.[8]

Max Perutz and the phase problem

The problem that needed solving in order to progress from interatomic distances to individual atomic coordinates is known technically as the *phase problem*. Intensities of scattered X-rays, recorded photographically or by any other means, are always averages over the period of an electromagnetic oscillation: the differences in 'phase' between individual scattered rays are necessarily lost in the averaging process. But the phase differences are needed (in addition to intensity and direction of the scattered wave) if atomic coordinates are to be individually defined and the goal of the Medical Research Council group set up in 1947 at Cambridge University was to find a practical way to determine them. Max Perutz (born 1914 and still riding his bicycle back and forth to work as we write) directed the project and can be given the principal credit for its success.*

The history of Perutz's involvement is a fascinating one, a mirror of the ups and downs of wartime life in Europe. Perutz had come to the Cavendish Laboratory in Cambridge from Vienna in 1936, to work as a research student under J. D. Bernal. After the war began in 1939, he was fortunate to receive a Rockefeller grant to remain there, now as a research assistant to Lawrence Bragg. This grant was not only a means of support for Perutz himself, but also provided the financial security that made it possible for Perutz's parents to seek asylum in England as refugees, which is what the entire family had indeed become when Austria was annexed by the Nazis in 1938.

---

* A personal account by Perutz containing reprints of essential papers, has been published recently.[9] Its emphasis is on the relation between the molecular structure of haemoglobin and the characteristics of its interaction with oxygen. An especially interesting aspect is Perutz's account of the battles he had to fight before most of his interpretations were generally accepted.

As it turned out, the war would soon affect Perutz in a less pleasant manner, threatening to disrupt his career at Cambridge before it got properly under way. After the Germans overran Holland and Belgium in May of 1939, British law decreed that all Germans and Austrians in Britain must be treated as 'enemy aliens', regardless of how implausible that label might be in the case of someone who was himself a refugee from the same enemy. The law called for arrest and internment: Perutz (with many others) was interned on the Isle of Man and, after a couple of months, transported even further away to Canada. Fortunately, the absurdity of this situation eventually became apparent, and Perutz was able to return to Cambridge in January 1941. A little later he was drafted for work in support of the war, joining 'Project Habbakuk', the Anglo-American project intended to create an airfield in the mid-Atlantic, to be constructed from ice and woodpulp—a project that had J. D. Bernal as the principal scientist.[10] Even then, a further slight hitch occurred because Perutz, though released from internment, was still officially classified as an 'enemy alien'. On his way to Washington for a joint US/UK planning meeting on the ice floe project, he was, not surprisingly, refused entry to the US. This difficulty was resolved when Perutz was speedily granted British citizenship by naturalization, a citizenship that he has retained ever since.

'Project Habbakuk' came to nothing in the end and Perutz returned to Cambridge and his haemoglobin crystals. He was joined in 1946 by John Kendrew (1917–1997), a physical chemistry graduate of Cambridge's Trinity College. In 1947, Perutz was made head of the newly constituted Medical Research Council Unit for Molecular Biology—with John Kendrew at first representing its entire professional staff! The way to cope with the phase problem followed slowly but surely after that. The basis was a fundamental and well-known fact, namely that heavy atoms have a much higher scattering intensity than lighter atoms. Crowfoot and Bernal had already half-heartedly looked at insulin crystals in which the native Zn atom of the protein had been replaced by the heavier Cd atom, but understandably the presence of one heavy atom among thousands has by itself no significant effect. What Perutz did was to apply a method beyond simple substitution, the method of *isomorphous replacement*, involving a search for situations in which incorporation of a very heavy atom would leave the protein structure and the structure of protein crystals otherwise unchanged. Therefore the diffraction *pattern*, the distribution of spots on the photographic image, would also remain unchanged. Only the intensities of individual spots would be affected: in a Patterson map, for example, which, as we have said, represented interatomic distances, every spot that involved the heavy atom replacement would become more intense—and thereby distinguishable from the rest. It was a difficult method, because in practice several independent heavy atom replacements at different positions were desirable. When successful, however, side-by-side comparison between native and the several modified proteins put elucidation of the three-dimensional structures of even complex proteins within the realm of possibility. Needless to say, the necessary computing effort was staggering in magnitude. (Data were still being fed by means of punched cards into the machinery of the embryonic computers that were available.)[11]*

Haemoglobin and myoglobin

While Perutz continued to focus on haemoglobin, John Kendrew began an independent project on myoglobin, similar to haemoglobin in being an oxygen carrier with the same iron-haem prosthetic group at its oxygen binding site, but also simpler because it was only a quarter the size of haemoglobin: one polypeptide chain instead of four.†

Because of its smaller size, the myoglobin structure was the first to be

---

* Lawrence Bragg, director of the Cavendish Laboratory, maintained enthusiastic support for the protein project through all the years of struggle. Francis Crick was among the many other famous people who became members of the Cavendish's molecular biology group. Bragg's blessing for these excursions into biology was by no means a foregone conclusion, for the historical fame of the Cavendish Laboratory rested on the achievements of some of the world's greatest 'pure' physicists—for example, James Clerk Maxwell, Ernest Rutherford, William Bragg. It required considerable imagination to foresee the huge influence on biology that would result from the formal marriage between physics and biology that the designation 'molecular biology' came to represent.[12]

† Myoglobin's function is to store oxygen (originally supplied by haemoglobin) in the tissues. It is particularly important for diving animals, such as whales, seals, and penguins, which need huge amounts of it to satisfy their needs during periods when atmospheric oxygen is not available to them. This explains the rather esoteric choice of the sperm whale as source for the protein that was used for the X-ray analysis.

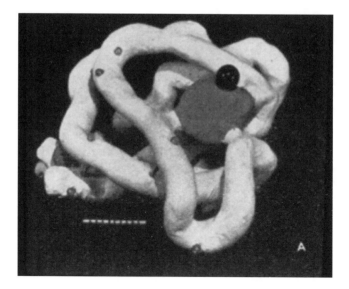

Fig. 13.1 Three-dimensional structure of myoglobin at low resolution (1958). (Source: Reprinted with permission from *Nature*.[13] Copyright 1958 Macmillan Magazines Ltd.)

completed. The first actual published model (in 1958) was based on data that could give a resolution no better than 6 or 7 Å, and, as shown in Fig. 13.1, that is not good enough to define exact atomic positions.

The structure of a protein molecule at 6-Å resolution is not unlike viewing an out-of-focus photograph in which dense objects, such as people, are seen as dark blurs against a background of less dense greys and whites. Thus, in Fig. 13.1, one can see electron-dense regions where the constituent amino acids are close together and with some imagination can trace the polypeptide backbone and close packing of amino acid residues. This may not seem a particularly startling result, but Kendrew's picture was newsworthy and in a sense surprising since many people had anticipated the appearance of some prominent feature: something like a regular packing of α-helices. In Kendrew's own words:

> Perhaps the most remarkable features of the molecule are its complexity and its lack of symmetry. The arrangement seems to be almost totally lacking in the kind of regularities which one instinctively anticipates, and it is more complicated than has been predicted by any theory of protein structure.[13]

And, in the same paper: 'We are [still] ignorant of the secondary structure of any globular protein. True, there is suggestive evidence ... that α-helices occur in globular proteins,' but even this could not be considered as proved and could not be gauged quantitatively. Eight heavy atom derivatives are listed as having been employed ('among others') for the analysis, combining with the protein at five distinct sites. The reagents used were mostly complex ions (for example, mercury iodide), those chosen being demonstrably restricted to single specific binding sites.

By extending intensity measurements to more and more spots, nearer the periphery of the diffraction map, ever shorter interatomic distances became available and the requisite data for achieving 2-Å resolution were obtained by 1960, with significantly greater effort. The number of reflections needed for 6-Å resolution had been 400, but the number became 9600 for 2-Å resolution. The intensity for each had to be measured separately for the unsubstituted protein and for each heavy atom derivative, and all had to be incorporated into the massive computer program that was used to analyse the results.[14]

Even at 2-Å resolution a frustrating difficulty emerged. The determination of the amino acid sequence for the polypeptide chain of myoglobin had not yet been completed. Without knowledge of this sequence, there remained ambiguities in parts of the structure, which would have been solved by steric restrictions if one had been certain about which particular side chains had to be fitted in. The response to this problem from Kendrew *et al.* was to aim for even higher resolution from their X-ray data alone—with high enough resolution, a sequence determined by chemical means might become unnecessary?

And so it turned out. With data allowing 1.5-Å resolution (20 000 individual reflections), the structure was solved. This was surely the icing on the cake! The amino acid sequence could be deduced from the shapes of the maxima in the electron density map that corresponded to side-chain atoms—that is, from spots that represented purely physical intensities without explicit chemical identification.[15-17]

The structural analysis of haemoglobin crystals was, of course, progressing in parallel; horse haemoglobin was used at first because its crystals had desirable characteristics, but the studies later shifted to the human species. The protein consists of four subunits, each superficially similar to the myoglobin molecule. The polypeptides are of two kinds, conventionally labelled alpha and beta. A low-resolution structure was

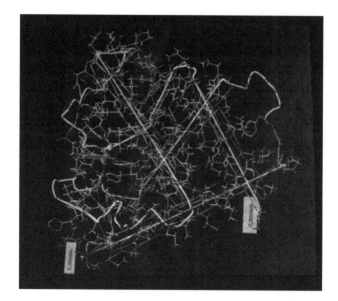

Fig. 13.2 Three-dimensional structure of myoglobin at high resolution (1961).
(Source: Reprinted with permission from *Nature*.[15] Copyright 1961 Macmillan
Magazines Ltd.)

published in 1960, simultaneously with the 2-Å result for myoglobin.[18,19]
But about ten years would elapse before one could even begin to see
details comparable to those that were provided for myoglobin.

Part of the reason for the delay can be found in the physiological function of haemoglobin, which is more complex than that of myoglobin.[20]
Measurements of oxygen binding had shown many years earlier that
binding did not obey thermodynamic equations for association of a
ligand with four independent sites on the same molecule, but indicated
instead a strong cooperative interaction between the four sites. Thus
deoxy-Hb and oxy-Hb were expected to have significantly different
structures, which needed to be determined separately.[21] The added
complication of the slow oxidation of oxyhaemoglobin to methaemoglobin (iron atom in the ferric state) made the determination of these
structures a formidable task. However, with myoglobin as a guide to the
folding of individual α and β chains, 2.8-Å resolution was enough to
construct at least tentative atomic models, and the earliest such structures (for oxy- and deoxyhaemoglobin from the horse) were published in
1968–1970.[22,23] The first theories to explain the special features of oxygen
binding (for example, the cooperativity) in terms of structural differ-

ences between oxy and deoxy forms were published about the same time.[24]

But a structure refined to the ultimate level of resolution (1.74 Å), where the position of every atom could be deduced directly from the X-ray data alone, did not become possible until 1984—dramatic technical improvements (synchrotron radiation) were required to achieve it.[25]

Excellent reviews are available,[26,27] some at a relatively popular level, which serve to summarize progress towards the detailed picture that ultimately emerged, defining the difference between oxy and deoxy forms, such as changes in relative positions of the α and β subunits—changes that could be identified with the cooperative changes in binding affinity.

Significance 'beyond a particular protein'

The low-resolution data for haemoglobin were sufficient to show that each of the individual haemoglobin chains bore a striking resemblance to the myoglobin structure.[28] Similarly, seal myoglobin molecules were found to be similar to those of sperm whale myoglobin, in spite of sub-stantial differences in the way the molecules were arranged in the crystals. It became apparent that the result obtained for the high-resolution structure of myoglobin represented 'a structure the significance of which extends beyond a particular species and even beyond a particular pro-tein'.

The end of this quotation is noteworthy. From everything we have related in the three preceding chapters, it is evident that the folding of proteins must follow rules set by physics—geometrical constraints, energy minimization, etc. A variety of rules had been suggested, some based on quite rigorous theoretical reasoning, others more speculative; sometimes they were mutually contradictory, as might be expected from the variety of sources. The broader significance of the myoglobin structure is that in effect it tells us what those rules are *for all proteins*. The picture seen in Fig. 13.2 resolved at one stroke many of the doubts and disputes that were still lingering in the protein literature.

Thus Pauling's α-helix was decisively confirmed by the exact dimen-sions of the helical segments within the myoglobin molecule. On the other hand, Pauling's further contention that hydrogen bonds between amino acid side chains are responsible for folding the entire molecule into a uniquely defined configuration was proved to be invalid—non-polar parts of the molecule are predominantly in the interior and charged groups are exclusively on the surface, making hydrogen bonds to water and not to each other. The point is that the 'rules' underlying these observations must surely be the same for all proteins—the Scheraga model for ribonuclease, designed to maximize hydrogen bonds between charged side chains (cf. p. 140), was, for example, effectively demolished by the myoglobin structure, even though the two proteins are quite unrelated. Likewise the myoglobin structure was used as a testing ground for spectroscopic tools that were being developed to suggest bits and pieces of molecular arrangement—for example, 'percent α-helix' in the folding of a polypeptide chain, based on measurement of circular dichroism or optical rotation.[29]

And the firmer the rules, the firmer became the notion that three-dimensional structure of a protein is determined by the primary sequence of a protein's amino acids. Even prediction of structure from amino sequence alone became a not impossible goal, though it is not yet realized in practice.

This was the apogee of the quest to solve the mystery of protein structure, appropriately recognized in 1962 by the Nobel Prize for chemistry, shared by Perutz and Kendrew.*

The discovery of the DNA double helix was honoured in the same year by the Nobel Prize for physiology and medicine, given to J. D. Watson, Francis Crick, and Maurice Wilkins, two of whom were colleagues of Perutz and Kendrew at the Cavendish laboratory. It was an unprecedented triumph for a single laboratory, never equalled since then.

Concluding comment

Since myoglobin and haemoglobin, three-dimensional structures have mushroomed. The first enzyme structure, that of lysozyme, appeared in

---

* Without any question, credit for this historical milestone belongs to Perutz and Kendrew and their co-workers at Cambridge, and to no one else. Pauling's 1951 set of papers on the α-helix and β-sheet pale into insignificance by comparison: they involved no study of protein crystals and provided no means for even guessing at the structure of a real protein molecule. The adulation sometimes given to Pauling in relation to this subject—in particular the claim by his biographer (T. Hager) that he came very close to being 'the first man to describe the structure of a protein'[30]—is obviously unjustified.

1962, and was quickly followed by high-resolution data, which allowed identification of the cleft into which its substrate fitted—the first pictorial demonstration of the 'lock-and-key' analogy that so many people, from Emil Fischer onwards, had invoked.[31–34] And, 36 years after initiating single crystal studies with pepsin as a graduate student, and 33 years after her first look at insulin, Dorothy Hodgkin returned to protein crystallography and obtained the three-dimensional structure of insulin at 1.9-Å resolution.[35,36] The total number of completed structures is today of the order of several thousand.

PART III

Physiological function

CHAPTER 14

An ancient and many-sided science

The exact and definite determination of life phenomena which are common to plants and animals is only one side of the physiological problem of today. The other side is the construction of a mental picture of the constitution of living matter from these general qualities. In this portion of our work we need the aid of physical chemistry.

Jacques Loeb, 1897[1]

Philosophical introduction

The dictionary defines 'physiology' as the science of the process of life, in both animals and plants, and today we take it for granted that proteins control all parts of this process and can provide the structural virtuosity to match the needs of even the most intricate tasks. Indeed, we shall see in several examples below that the die-hards, the non-reductionists, who ignored the growing trend to explain function in terms of protein structure, were quickly left out in the cold. But we must also appreciate that the science of 'the process of life' is much older than protein chemistry, going back to before proteins were known at all. Writing about the history of protein involvement inevitably takes us back to a time when the ideal of molecular explanations for physiological function could not have existed.

Much publicity has been given to vitalism in this context, the doctrine that the laws of physics and chemistry could never by themselves explain the phenomena of life at all, that explanations of the kind that governed engines and other inorganic aspects of an increasingly mechanized society could not be expected to account for the complexities of biological processes. Surely this doctrine was defunct by 1900, one would think, but the evidence is otherwise, even for some who espoused the role

Fig. 14.1 Jacques Loeb in 1895: advocate of the use of physical chemistry to explain physiological function. (Source: P. J. Pauly (1987). *Controlling life*, Oxford University Press.)

of proteins in physiological function. As late as 1906, the introduction to a well-respected textbook of protein chemistry reads as follows:

> Biologists of the present day may be divided into those who believe all animal and vegetable existence to be endowed with some special unexplainable force, called the 'vital principle', and, on the other hand, the physico-chemical school who endeavour to expound organic life by only those laws which hold good for the lower inorganic compounds.

> The view the author holds, he trusts, will bridge over the gulf existing between the two schools mentioned above, namely, those of the vitalists and the non-vitalists.[2]

Needless to say, this gulf could not be bridged: there can be no common ground between protein science and any doctrine that embodies an 'unexplainable' force.

However, another philosophical group must be taken more seriously as creating hurdles on the path to the recognition of proteins as the basis for physiological processes. They were the natural philosophers, most of them German, greatly admired and respected, who considered themselves the antithesis of vitalists and who, over a period of a century or more, espoused strictly rational, tangible approaches to the secrets of nature. Their fault, as we perceive it today, was that the rationality came from within the philosopher's head, with no hard training in physics or chemistry required—possibly a partial explanation for their popularity, for what could be more awesome than the ability to utter profundities about nature without getting your hands dirty in the process.

Immanuel Kant (1724–1804) was an example, author of the intellectually brilliant *Critique of pure reason*.*

Another was Johann Wolfgang Goethe (1749–1832), the famous poet and dramatist, who was fully accepted in Germany as a respectable scientist.[4,5] His most famous scientific work, *Zur Farbenlehre* (1810), dealt with optics and colour. It was based on the conviction that Newton was wrong and a charlatan, and proceeded to elaborate a theory, splendid in its poetic intuition, that could have been disproved by the simple fact of the existence of a rainbow.

Goethe was not unique. Being a poet, he may have been allowed greater licence than some, but natural philosophers in general had pretty free rein to say what they wanted. Philosophers like Ernst Mach in Vienna questioned the existence of atoms, the very core of all thinking about protein structure. As late as 1900, the great physicist Ludwig Boltzmann, much of whose career was devoted to proving the existence of atoms, was

---

* Kant's work and influence are difficult to understand in an age committed to experimental science. The subject is discussed in virtually every encyclopaedic reference book. The essence is always the same; for example, we cite here an entry from the *Oxford dictionary of philosophy*:[3]

> In spite of the notorious difficulty of reading Kant, made worse by his penchant for scholastic systematization and obscure terminology, his place as *the greatest philosopher of the last three hundred years* is well assured.

So be it: we must take the philosopher's word for it. (The italics are ours.)

driven to distraction by Mach's persistent objection: 'Who has ever seen one?'

The point to be made is that anybody setting out to do research on biological processes would not have been entering virgin territory, but would often be faced with entrenched ideas that came from a different world, a non-chemical world, ideas that by his own standards could not withstand much scientific scrutiny. Hermann Helmholtz (1821–1894), one of the most illustrious pioneers at the interface between physical science and physiology, had inherited a profound concern about music, art, and philosophy from his father, who particularly admired Immanuel Kant and J. G. Fichte (1762–1814), a fervent Kant disciple.[6] When he first introduced his theory of visual perception, he felt the need to acknowledge his admiration for Goethe's philosophies at the same time as he refuted the specifics of Goethe's optical theories.[7] Helmholtz's most famous early lecture on vision, 'Über das Sehen des Menchen', was given (in 1855) at a meeting honouring Immanuel Kant, and Helmholtz had to apologize that his lecture was perhaps inappropriate, his empiricist theory not at all in the Kantian spirit.[8]

The limitations of pure philosophy were eventually swept away and new spirits appeared on the scene, such as Jacques Loeb, enthusiastic champion of modernity and a mechanistic point of view.[9] But there was still considerable kinship between Loeb and the natural philososophers, in their manner of thinking. Loeb, too, tended to express a general philosophy instead of coming to grips with specific problems. His thesis was thoroughly 'modern' in that physical chemistry was what he advocated, but his explanation of the need for physical chemistry was vague in as to exactly how physical chemistry would be applied. To quote directly from an 1898 polemic: 'we need the aid of physical chemistry and especially of three of its theories: stereochemistry, van't Hoff's theory of osmotic pressure and the theory of the dissociation of electrolytes.'[1] That's more a definition of 'physical chemistry' than a recipe for research. There was no indication that Loeb (at that time) saw molecular specificity as the basis for functional diversity, as the target to which physical chemistry must be aimed.

Considerably later, in a somewhat similar vein, we have A. V. Hill, physiology professor at Cambridge, a brilliant man by all accounts, and unquestionably a thermodynamicist par excellence. He knew about proteins, but *chose* to ignore them. He devoted a lifetime to muscle research

(Chapter 18), but firmly believed that proteins were the province of biochemists, not his concern as a physiologist.[10]

Brewers and physicians

Medical science provides other examples of protein chemists arriving late upon the scene: here they enter upon ground prepared often centuries before by physicians and their sometimes ancient cures and nostrums. Their contemporaries in the medical field favoured practical solutions over molecular explanations: the idea that protein chemistry could be the source of new benefits for medical practice would not have been common in the early days.

The relationship between diabetes and the protein insulin is a perfect example. Diabetes had been known for centuries—the 'pissing disease' it was called, characterized by unquenchable thirst and constant flow of urine. In this case the cure of the disease, the discovery of insulin at the University of Toronto in 1921–1922, was a dramatic event, one of the genuine miracles of modern medicine.[11] But it all occurred ahead of any knowledge of insulin as a protein; more than that, the fact that insulin is chemically a protein was effectively irrelevant to the discovery. The beneficiary here has been protein chemistry, in that the huge quantities of the protein that were required for clinical use guaranteed availability for research as well.

Immunity, too, was recognized long ago and 'antisera' were in use long before 'antibodies' were purified from blood serum. In this case, however, philosophers had little to say on the subject and even medical knowledge at the chemical level was primitive until protein science came into play. In fact, it is fair to say that, until Tiselius isolated the γ-globulin fraction of blood serum in 1935, no understanding of immunity on the molecular level would have been possible. Since then the relationship between medicinal use and chemical science has been symbiotic, progress in either one triggering progress in the other. In a sense, benefits have also been divided both ways: the remarkable specificity of antibodies and the ability to 'culture' them in the blood, so to speak, as guardians against specific diseases, has also permitted them to be 'cultured' in the same way as analytical reagents, specifically active against proteins unrelated to disease.

A still different kind of example is provided by commerce, by the brewing of beer and the making of wine. There was no primary academic interest in this case, nor even the altruistic motivation of some medical doctors. And no philosophy, other than that attending the never-ceasing debate about use or abuse of alcoholic beverages. But it is interesting to note that protein science has actually been well served by what were initially purely commercial aims. The best example is the Carslberg brewery in Copenhagen: Jacob Jacobsen (1811–1887), its owner, was a prescient man, who could see the potential benefits of science for his business and who founded the famous Carlsberg Laboratory in 1876, on the grounds of the brewery, where it remains to this day.[12] In this case it was definitely of mutual benefit and we have mentioned Carlsberg repeatedly in earlier chapters for its direct contributions to the understanding of protein chemistry, especially in Chapter 5. And in more general terms, where would enzymology be without the brewers?

Scope of our review

To reiterate what we stated at the beginning of this chapter, we take it for granted today that proteins control all physiological function. The molecular virtuosity required for this all-encompassing mission is huge. How can a few hundred amino acids, strung together in one or a few polypeptide chains, carry out the virtually infinite variety of tasks that physiologists have described? How long did it take in individual cases to answer this question and to define the molecular mechanism? Was there a period when the central role of proteins was not appreciated or even vigorously opposed?

These questions define the scope of this section of our book. The huge diversity of physiological function means that we must limit ourselves to just a small sample of the rich structure/function literature that is available, cameos, if you like, of parts of the overall historical heritage.

Enzymes, of course, take pride of place: recognition and subsequent understanding of the numerous specific enzymatic catalysts is a subject that is virtually synonymous with the total history of biochemistry.[13] Here we concentrate on the recognition of the chemical nature of these powerful catalysts, the identification with proteins being surprisingly resisted until well into the twentieth century.

A chapter on antibodies turns our attention to preservation of the species against ever-present attack from a hostile environment. Still chemistry, but how can the system be programmed in advance for whatever enemy it may have to confront in the future? Or perhaps it is not programmed in advance, but the protein structure is adaptable, capable of being modified by the enemy as it prepares to strike?

A short chapter on rhodopsin and related proteins deals with the most highly prized of human sensory perceptions, colour vision. If poets are to have their say, here would be the place.

Muscle contraction focuses on what has through the ages been thought to be the defining characteristic of animals (as distinct from plants)—the power of independent movement. Here the few hundred amino acids in a chain actually produce motion: one polypeptide chain pulls on another and scores of them acting in concert pull an arm or a leg. Chemistry has been converted to mechanics, but the intrinsic definition of catalysis still applies: the protein is not changed in the process (as Liebig and others thought it would be) and can be reused over and over again.

We conclude with a chapter on the proteins of cell membranes. It can hardly be called a 'history', because the topic is too modern, details still not completely worked out. But the individual cell and its membrane are the ultimate element of all life—proteins, as expected, are in charge at the interface between the cell and its environment, controlling traffic back and forth of food, energy, and refuse.

We have omitted as much as we have included; for example, here are some topics where protein structure and physiological function have been just about as closely related as in the topics we have chosen for explicit discussion.

Allosteric enzymes. Enzymes take pride of place, as we have said, and understanding of specific enzymatic catalysis is virtually synonymous with all of biochemistry. Allosteric enzymes illustrate the progress in sophistication that took place in the 1960s, taking us from the simplicity of early lock-and-key concepts, as illustrated in Chapter 15, to a more dynamic picture, to changes that can modify the catalytic process and thereby regulate it to meet physiological requirements. Historically, the subject is closely related to the cooperative association between oxygen and haemoglobin (see Chapter 13) which *increases* the binding affinity for oxygen as the degree of saturation increases.

The most enthusiastic proponents of the idea were the French molecular biologist Jacques Monod and his colleagues at the Institut

Pasteur in Paris,[14,15] who thought they had discovered one of the great universal principles of biochemistry: they called allosteric proteins 'the most elaborate products of evolution' and occasionally re-invented some already established protein chemistry (mostly related to subunits) in the course of their research into the subject. The modern view, while not denying the importance of the formal allosteric mechanism, recognizes it as only one of several regulatory schemes, many of which do not involve conformational change of the enzyme.

Nevertheless, the initially exaggerated excitement served a useful function in protein science as a whole. Molecular structures as revealed by X-ray crystallography ceased to be regarded as immutable—more attention was paid to alterations in conformation that could happen as a result of interactions. The classic account of the theory of ligand-induced structural change is that of J. Wyman[16] and there is a modern historical (and philosophical) account by Creagar and Gaudillière.[17]

Blood clotting. Physiologically, the purpose is to seal the wound and hold in the blood. A genetic defect leads to absence of one of the proteins, with dramatic effects on the histories of the English and Russian royal families.[18] Chemically, the process involves a total of 13 or 14 plasma proteins, at least one tissue protein, calcium ions, membrane surfaces, and platelets. Since about 1960 the components of the system have been known as numbered factors: factor I is fibrinogen, factor II is prothrombin, and for most of the rest the numerals themselves are usually used in common parlance. Factor VIII is the missing activity that leads to classical haemophilia.

Many of the steps in the clotting mechanism involve a cascade of proteins that is generated by means of successive proteolytic reactions:[19,20] it is the most elaborate known example of the precursor principle that was illustrated in Chapter 9 by the relation between insulin and proinsulin.[21]

Oxidation–reduction: the cytochromes. Physiologically, the problem was the pathway by which energy is derived from the reduction of oxygen: what are the intermediates between O_2 and CO_2? Chemically, the quest led to the cytochromes, successive parking places for electrons in the scheme of bioenergetics.[22] The principle of the action of cytochromes is not unlike the principle involved in colour vision, the function of the protein being to modify the properties of a bound prosthetic group— in vision it modifies the absorption spectrum; here it also affects the absorption spectrum, but, more importantly, it modifies the oxidation/reduction potential. An added interest is the universality of the system: it is common to all organisms that follow the aerobic mode of life.

Concluding comment

One could go on *ad infinitum*: there is no limit to the adaptability of proteins. Those who seek a sociological explanation for the driving force behind the often evangelical enthusiasm of protein researchers need go no further than this. Proteins can do for you whatever you ask of them.

CHAPTER 15

Are enzymes proteins?

The only question to be determined is whether that hypothesis is too bold which assumes that in the organism of yeasts there is a substance that decomposes sugar into alcohol and CO_2.

F. Hoppe-Seyler, 1881[1]

We have arrived, indeed, at a stage when, with a huge array of examples before us, it is logical to conclude that all metabolic tissue reactions are catalyzed by enzymes.

F. G. Hopkins, 1913[2]

In 1901 Franz Hofmeister, who was shortly to establish the peptide bond structure of proteins, predicted that in future a specific 'ferment' would be found for every vital reaction within the cell.[3] This view was restated unequivocally by F. Gowland Hopkins twelve years later—'all metabolic tissue reactions are catalyzed by enzymes'.[2] A long path had been travelled since 1752, when the French polymath, René-Antoine de Réaumur, first demonstrated the power of gastric juice extracted from the stomach of carnivorous birds to dissolve meat.[4]

The slow and contentious birth of enzyme science

Between Réaumur and Hopkins we had the nineteenth-century scientific wars over the mode of fermentation, which marked the real beginning of the science of enzymology and which have become an integral part of scientific folklore. There were those who espoused the 'cell theory', notably the Frenchman Charles Cagniard-Latour and the Germans, Theodor Schwann and Friedrich Kützing, and, a little later, the great popularizer, Louis Pasteur: their contention was that fermentation, the conversion of sugar into alcohol, *required* the presence of living yeast. On the other side there were the 'pure chemistry' advocates led by Justus Liebig, who denied any involvement of micro-organisms. And in a sense they were both right, but it took half a century to sort out the muddle.

Table 15.1 Nineteenth-century enzyme discoveries

Year	Discoverer(s)	Enzyme	Source
1833	Payen and Persoz[7]	Diastase	Malt extract
1836	Schwann[8]	Pepsin	Gastric juice
1837	Wöhler and Liebig[9]	Emulsin	Almond extract
1846	Bernard[10]	Lipase	Pancreatic juice
1860	Berthelot[11]	Invertase	Yeast extract
1876	Kühne[12]	Trypsin	Pancreatic juice
1895	Bertrand[13]	Laccase (oxidase)	Latex

During this period, it was not only the turning of sugar into alcohol that occupied would-be enzymologists. Secretions from many sources, both animal and plant, were found to influence the rates of some chemical reactions and this became a popular field of study. Even the chemist Jacob Berzelius comes into the picture, for it was he who coined the word 'catalysis' in 1836 to describe this action by which an agent affected a chemical reaction but was itself recovered unchanged afterwards.[5] It is not our intention here to discuss in detail all the nineteenth-century identifications of various types of enzymes, but some landmark discoveries are listed in Table 15.1. A thorough and well-documented history of the period—'from ferments to enzymes'—has been provided by J. S. Fruton.[6]

The substances listed in the table were extracts with catalytic activity, not in any sense purified, and the majority of them were hydrolytic agents seemingly secreted by living cells; in fact, most biologists thought at the time that only hydrolytic reactions were catalysed by substances secreted from micro-organisms. Laccase, discovered in 1895, was the first activity identified in the oxidation–reduction category.[13] All these catalytic substances were commonly referred to as 'unorganized ferments' to distinguish them from 'organized ferments' such as the intact yeast cell which many biologists held to be essential for the conversion of sugar to alcohol. It was Willy Kühne who suggested the term 'enzyme' in 1876, explicitly limiting the name to ferments that had been shown to function outside the cell.[14] Indeed, at that time, there were a large number of biologists who rejected the existence of intracellular enzymes. A notable exception to this was Felix Hoppe-Seyler, who postulated in 1881 that all intracellular chemical reactions were catalysed by the same type of enzymes that were found in cell secretions. However, he continued to hold the common belief that these reactions consisted solely of hydration and dehydration processes.[1]

The issue of the existence of intracellular enzymes should have been settled in 1897 when Eduard Buchner (1860–1917), working with his brother Hans (a bacteriologist), discovered zymase activity in the 'juice' extracted from yeast cells.[15,16] The Buchners were attempting to find a more efficient way of breaking open microbial cells in order to obtain extracts that might have medicinal value and had added sucrose as a preservative to the easily decomposed fluid that came out of the cells. It was pure serendipity that they noted the presence of something that tended to decompose the sucrose, leading to the identification of what we today know to be a mixture of several enzymes involved in the fermentation process. It was convincing proof that the living, intact yeast was not necessary (only the cell contents were required for catalysis of the chemical reactions involved) and soon hosts of physiological chemists were involved in studying metabolic processes in the test tube rather than the animal (or plant) body. Buchner received the Nobel Prize for chemistry for his discovery in 1907.*

Another thorny problem for researchers at this time was that of reversibility. If these substances were catalysts (in the sense that the word was then understood), they should promote reactions in both directions. The first such demonstration of reversibility came in 1898, when Arthur Croft Hill in England was able to synthesize maltose from two glucose molecules using the enzyme maltase.[17]

Specificity

It was obvious from the beginning that the enzyme catalysts of living organisms had specific substrates—those that hydrolysed sugars did not dissolve the albuminous materials found in cells, those that digested meat were ineffective against fats. However, the exquisite tailoring of enzyme to substrate became apparent only when Emil Fischer began his epochal

---

* Surprisingly, it was not a biologist but a chemist, Richard Willstätter, who was the most prominent die-hard, questioning what was a severe blow to vitalist notions and claiming as late as 1937 that Buchner's yeast juice behaved differently from intact yeast cells in the fermentation process. Willstätter will appear later in this chapter at the centre of an even more controversial dispute.

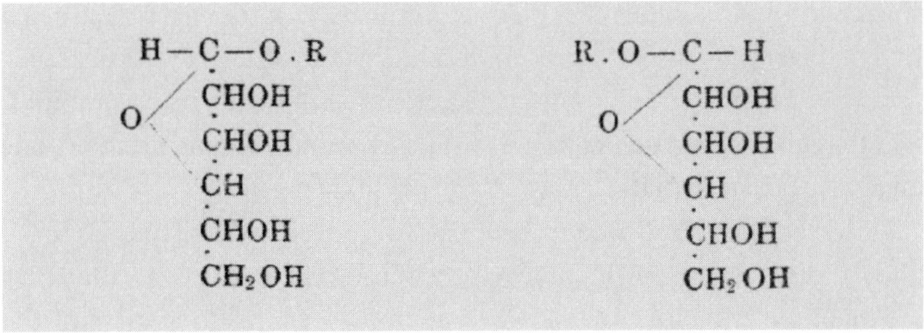

$$
\begin{array}{ll}
\mathrm{H-C-O\,.\,R} & \mathrm{R\,.\,O-C-H} \\
\quad\;\diagup\;\mathrm{CHOH} & \quad\;\diagup\;\mathrm{CHOH} \\
\mathrm{O} & \mathrm{O} \\
\quad\;\diagdown\;\mathrm{CHOH} & \quad\;\diagdown\;\mathrm{CHOH} \\
\qquad\;\mathrm{CH} & \qquad\;\mathrm{CH} \\
\qquad\;\mathrm{CHOH} & \qquad\;\mathrm{CHOH} \\
\qquad\;\mathrm{CH_2\,OH} & \qquad\;\mathrm{CH_2\,OH}
\end{array}
$$

Fig. 15.1 The formulas for α- and β-methylglucosides, as they appeared in Emil Fischer's 1894 paper. The only difference is the stereo-configuration about the C atom at the top end of the formulas, but the glycosidic enzymes distinguished them unerringly. What other than a protein could accomplish this feat? (Source: E. Fischer.[18])

investigations of glycoside hydrolysis in 1884. This renowned organic chemist had already established a formidable reputation and turned his interest in the stereochemistry of carbon-containing compounds quite naturally to sugars and their cleavage by the then known enzymes. Over the next fourteen years, his laboratory established unequivocally that enzymes were tailored to match specific stereoisomers of sugars. Invertase from yeast recognized and was able to split the synthetic compound α-methylglucoside, but did not catalyse the hydrolysis of the β form. The reverse result was obtained with the enzyme emulsin, and both results were confirmed by use of appropriate natural products which had the same stereochemistry. From these studies emerged Fischer's well-known 'lock-and-key' model for enzyme–substrate interactions; he used exactly those words in German ('Schloss und Schlüssel') in his original report.[18] The words of course imply a rigid fit between two substances, but the more modern view, which suggests a degree of conformational adaptability in the ability of enzymes to accomodate the substrate, was a logical outgrowth of Fischer's work, stimulated by more extensive knowledge of protein structure and its inherent subtleties.

Enzymes are proteins – or are they?

Until the early 1900s most physiological chemists had assumed that enzymes were either proteins or protein derivatives because the catalytic activity seemed invariably to reside in the water-soluble 'albuminous substances' contained in living cells.[19] When reporting on his studies of sugar hydrolysis, Emil Fischer himself speculated that the enzymes secreted by yeast were 'most likely protein in nature' and that, like the proteins, they have asymmetrically constructed molecules. Fischer's great ambition was that his polypeptide syntheses would eventually enable him to create an enzyme *de novo* in the laboratory. (See Chapter 3.)

However, the rise of colloid chemistry (discussed in detail in Chapter 4) and the discovery of pertinent instances of catalysis by inorganic materials led many investigators astray. When attempts were made to increase the purity of enzyme preparations, it was almost universally observed that enzyme activity did not co-purify with protein (as measured, say, by analysis of nitrogen or sulphur content). Lacking appreciation of the extremely high sensitivities of specific activity measurements, as compared to chemical tests designed for protein en masse, some scientists adopted the colloid chemist's concept of 'adsorption' and postulated that the proteins were merely carriers of the compounds actually responsible for the catalytic activity, the latter being adsorbed to the protein surface. Indirect support for this sort of picture came from knowledge of metallic surfaces that could adsorb various chemical entities, bringing them into close proximity for eventual reaction. Alternatively, one could point to the role of as simple a substance as hydrochloric acid as a catalyst in the hydrolysis of cane sugar and use it as an example of how an enzyme might work: HCl obeys the basic definition of a catalyst, greatly increasing the rate of reaction, but not being consumed in the process.

Evidence from closer to home came from the ability of haemoglobin to catalyse the oxidation of guaiaconic acid (a constituent of a gum resin), but it could not do so in the absence of iron, suggesting that the metal was the true catalytic factor.[20] Following on the heels of this experiment, Otto Warburg set up a number of what he called 'model systems' for oxidative catalysis, in which he showed, for example, that 'charcoal haemin'—iron adsorbed to the ash of burnt haemin—could be an effective catalyst for oxidation of some amino acids.[21] The ultimate demonstration for the colloid followers was undoubtedly the formation of an artificial enzyme from a mixture of gum arabic and manganese formate which functioned as a peroxidase—here there was no protein at all.[22]

In any case, there is no question about the fact that notable scientists of the period often felt uneasy about the association of enzyme activity with proteins: the two sides in the debate about enzymes remained evenly matched for a long time. Thus Otto Cohnheim (who had no doubts about the high molecular weights of proteins) was not at all convinced of the identity between enzymes and proteins: in lectures given in 1912 at Johns Hopkins University he said:

> We have no reason to think that all enzymes belong to the same class of chemical compounds. It has already been shown that the enzymes are accompanied as a rule by proteins and nucleic acids. The difficulties of separating enzymes and proteins have led physiologists for a long time to regard enzymes as protein-like bodies, though Brücke and others were successful years ago in freeing enzymes from all traces of proteins. . . . We know but little today about the chemical characteristics of enzymes.[23]

About the same time, Leonor Michaelis in Berlin was reasonably sure that invertase, trypsin and pepsin were all proteins and measured their isoelectric points as confidently as he measured the isoelectric point of serum albumin (cf. Chapter 5, p. 66).[24] Michaelis and his Canadian co-worker Maud Menten (1879–1960) went on to formulate a mathematical equation to describe the kinetics of enzyme action, based on the simplest possible theoretical model—reversible association between substrate and the enzyme's binding site, followed by unimolecular dissociation of the enzyme–substrate complex to yield product and regenerate free enzyme.[25] The equation remained for decades the standard tool used by biochemists for analysis of enzyme kinetics: it was extended to allow for the addition of inhibitors that would compete for the binding site, but not themselves undergo conversion; it was often used in reciprocal form to generate linear plots of experimental data from which numerical values for characteristic constants of the enzymatic process could be evaluated. (Studies of rates and equilibria by this equation did not, of course, by themselves require a decision about the chemical nature of the binding site.)

Carl Oppenheimer, in a greatly expanded 1913 edition of his enzyme textbook, sat on the fence. On the one hand, enzymes are certainly protein-like ('Eiweissähnlich'), he said, pointing out that they are amphoteric electrolytes and that low diffusibility is a common characteristic of all of them. But without a one-to-one chemical correspondence that proves nothing.[26]

The greatest enthusiasts for the concept of enzyme activity residing in an adsorbed small molecule were organic chemists, most famously Richard Willstätter (1872–1942), the German chemist, who had been awarded a Nobel Prize in 1915 for his work on plant pigments. Willstätter's interest in enzymes dated from the end of World War I when he and his colleagues embarked on an ambitious programme of purification. They developed more reliable assay methods and pioneered new adsorption techniques for separating components of cell extracts—these were important contributions to the field and led to the adoption of Willstätter's procedures by many biochemists and others.[27] But improved purification exacerbated the problem of defining the chemical nature of enzymes: specific catalytic activities tend to be so high that they could be measured using minute samples, which were far too small for applicability of quantitative chemical analyses that could define proteins as such. Ultimate purification reached the point where no protein could be detected at all by the assay procedures then in use.

This is the reason why ideas inspired by colloid chemistry had a much stronger influence on the question of whether enzymes are proteins than they did on the question of protein molecular weight, where the theory of colloidal aggregation was never taken seriously (see Chapter 4). In the case of molecular weight, every improvement in purification led to better and better confidence in intrinsically macromolecular proteins, with precisely definable molecular weights. In the case of enzyme catalysis, improved purification seemed to lead towards total disappearance of the protein as a functional component. How could one dispute the colloid chemist's speculation, the idea that enzymes, when found associated with proteins, were simply small organic/inorganic molecules non-specifically adsorbed on the surface of the ubiquitous protein molecules? Support for this idea from a scientist of Willstätter's stature had a strong (and negative) effect on the search for the chemical nature of enzymes.

James B. Sumner and the crystallization of urease

It was only after 1926 that the protein nature of enzymes was experimentally and unequivocally demonstrated, following the first crystallization of an enzyme, urease, derived from the jack bean plant, by James B. Sumner at Cornell University in Ithaca, NY. The event is regarded in the annals of protein science as close to equal in importance to the recognition of the peptide bond as the structural basis for proteins. (From now on the number of distinct proteins would rise by leaps and bounds: every biochemical reaction needed a unique enzyme; every enzyme had to be a distinct protein.)

James Batcheller Sumner (1887–1955) was a young Ph.D. from Harvard when he entered upon his first academic position at Cornell University, doing so at about the same time that Willstätter moved into the enzyme field. Sumner was endowed with more than the usual amount of stubborness and determination as evidenced by his life both before and after his historic achievement and the subsequent award of a Nobel Prize. He had lost the use of his left arm in a shooting accident as a young boy, a serious enough handicap for anyone but more so for Sumner since he was left-handed. He was advised that it was well-nigh impossible for him to consider chemical research as a career—a challenge that he met with great vigour, proving himself more than capable at the laboratory bench. As it turned out, equal determination was needed to gain acceptance of the results of his research, against the entrenched beliefs of Willstätter and his followers.[28]

Sumner's appointment at Cornell involved two lecture and laboratory courses in biochemistry to medical and home economics students, together with two advanced courses and a seminar. All this with the aid of just one graduate assistant. Equipment for research was meagre and funds in short supply. Many years later in his Nobel Lecture (1946) he said:

> I wish to tell next why I decided in 1917 to attempt to isolate an enzyme. At that time I had little time for research, not much apparatus, research money or assistance. I desired to accomplish something of real importance. In other words, I decided to take a 'long shot'. A number of persons advised me that my attempt to isolate an enzyme was foolish, but this advice made me feel all the more certain that if successful the quest would be worthwhile.[29]

Since Sumner was convinced that enzymes were proteins and because he was already familiar with jack bean urease as an enzyme from his Ph.D. research, his protocol began with the separation of all the 'globulins' in the plant seed, followed by successive fractionations on the basis of urease activity. It was to become a nine-year programme, culminating in a short paper in 1926 describing the purification and crystallization of urease, documenting the positive tests for protein and the negative tests for other chemical substances such as carbohydrate and fats which were possible contaminants in his preparation.[30] Working essentially alone and unaided, the amount of crystalline urease he was able to prepare at any one time was extremely small, so it required another six years to

Fig. 15.2 James B. Sumner: the first to crystallize an enzyme protein. (Source: © The Nobel Foundation)

produce eighteen papers further characterizing his crystals. He subjected his preparations to inactivation studies, isoelectric point measurements, antibody reactions, and digestion by proteolytic enzymes to show concomitant loss of activity and protein structure. Nevertheless, over these last six years, the 'carrier theory' of Willstätter and his followers continued to dominate conservative opinion. The continued hostility to the idea that enzymes are proteins, lasting into the late 1920s, is difficult to understand today.[31]

Fig. 15.3 John H. Northrop: crystallized pepsin in 1930, four years after Sumner's urease. (Source: © The Nobel Foundation.)

John H. Northrop and the crystallization of pepsin

In contrast to James Sumner who began his independent career on a high-risk project carried out in less than auspicious surroundings, John Howard Northrop (1891–1987) embarked on the purification and crystallization of pepsin as a well-established investigator at the Rockefeller Institute—no onerous teaching duties, no funding problems, surrounded by other scientists to whom he could turn for aid and advice. He had worked with Jacques Loeb studying the kinetics of various enzymatic processes and had even been involved with the latter's 'theories of the duration of life'.[33] In 1930 he published the seminal papers that were to earn him a share of the Nobel Prize (with J. B. Sumner and Wendell Stanley) sixteen years later.[34,35]

Also unlike Sumner, Northrop was able to begin his work with an enzyme that was already partially purified, a crude commercial preparation of pepsin, which he was able to purchase in large quantities and which at the end of his labours produced an incredible 2 kg of crystalline material. Consequently, his first paper establishing the protein nature of pepsin contained every conceivable test that was possible at the time—solubility studies, inactivation by heat and acid, measurement of diffusion coefficient (yielding a molecular weight of 37 000 when combined with the assumption of a spherical particle with the density of a protein), antibody inhibition of activity (acknowledging the advice and assistance of two of his colleagues at the Institute).

Northrop's own publications add to the evidence that it was not until after the crystallization of pepsin (and later purifications of trypsin and chymotrypsin) in his own laboratory that the concept of an enzyme as a protein became widely accepted. Northrop himself was certainly familiar with Sumner's work of 1926, but his evaluation of it was at best sceptical, as evidenced by his comments on his own discovery: 'There seems to be no convincing evidence that any enzyme has been obtained in the pure state; and only one, the urease described by Sumner, has been previously obtained in crystalline form.' Similarly, in 1935 when reviewing the chemistry of pepsin and trypsin, Northrop stated: 'In the meantime, Sumner (1926) reported the isolation of a crystalline protein from beans which *appears* to be the enzyme urease.'[35] Elsewhere he conveys the message that it is not necessarily true that *all* enzymes are proteins.

Willstätter and his followers greeted both Sumner's and Northrop's work with frank disbelief and even when they were forced to concede that the 'protein carrier' of the adsorbed enzymatic activity might in some way affect the chemical process involved, their adherence to 'colloid dogma' continued. A particularly lively interchange took place in the journal *Science* in 1933. E. Waldschmidt-Leitz, one of Willstätter's closest adherents, concluded a short dissertation on the chemical nature of enzymes with the statement:

> These crystalline enzyme preparations should be regarded, then, as adsorption compounds of the true enzymatic component plus crystalline protein to which they have a special affinity. The finding of crystalline

protein–enzyme compounds may lead to the concept that enzymes are merely proteins, and thus cause investigators to disregard enzyme specificity which can only be explained by the existence of highly specialized active groups.[36]

Two weeks later Sumner replied quoting the above paragraph.

I think there is little danger of this. The enzyme, as I consider it, is in some cases a simple protein, in others a conjugated protein where the properties are to be ascribed to the molecule as a whole. But whether the specific active groups are in the protein part or in the side chain, the enzyme is a protein, as I demonstrated in 1926.[37]

Epilogue

James Sumner continued to purify proteins from the jack bean and in 1937 purified and crystallized catalase from this source. John Northrop and Moses Kunitz isolated the zymogen precursors of pepsin, trypsin, and chymotrypsin[35] and made many other contributions to the field of enzyme kinetics. Richard Willstätter made yet another career change at the end of the 1930s when the protein nature of enzymes was finally almost universally accepted. He embarked on studies of the biochemical transformation of glycogen and the role of polysaccharides in alcoholic fermentation. In the words of his biographer, 'This work did not have a significant impact.'[38]

CHAPTER 16

Antibodies

> To attribute what could be called inventive activity to the body or to its cells, enabling them to produce new groups of atoms as required, would involve a return to the concepts current in the days of [an obsolete] natural philosophy. Our knowledge of cell function and especially of synthetic processes would lead us rather to assume that in the formation of antibodies, we are dealing with the enhancement of a normal cell function, and not with the creation at need of new groups of atoms. Physiological analogues of the group of the specifically combining antibodies must exist beforehand in the organism or in its cells.
>
> Paul Ehrlich, 1897[1]

The knowledge that there is such a thing as immunity goes back to the Greeks and even earlier.[2] Some people died from disease and others survived; the latter were sometimes immune to any subsequent infection. Then, in 1798, Edward Jenner showed that immunity against smallpox could be generated by the infectious agent that led to the related but less virulent cowpox—the process we call vaccination. It was one of the most famous milestones in the history of medicine, the root of the string of discoveries that have made headlines since then and continue to do so to this day.

Understanding immunity was, of course, never confined to the identification and characterization of immunity-conferring proteins. Just as attempts to understand the process of fermentation led to conflicts between scientists who espoused the need for the presence of a living yeast cell and those who claimed that pure test-tube chemistry provided all the means to convert sugar into alcohol, so too the investigation of immunity produced a division between those who looked to cells for an answer and those who sought single chemical entities. In the case of fermentation, both groups were right in a sense—the living cell was often the active agent in commercial processes, but the cell could be lysed to produce enzymes that could do the same job outside the cell by means of catalysis of what turned out to be perfectly sensible chemical reactions.

In the field of immunology the problem was not so easy. There were both molecular and cellular aspects, but the time scale required for the two to merge to a point where they could be related was much longer and the relationship itself more complex than in the case of enzymes. Cellular aspects languished in the wilderness for many years while chemical studies of the active molecule (the immunoglobulin protein) were progressing rapidly.

Cellular and humoral immunity

By the second half of the nineteenth century, thanks to the efforts of Edward Jenner (1749–1823), Louis Pasteur (1822–1895), Robert Koch (1843–1910), and others, the germ theory of disease was widely accepted and vaccination by either attenuated or dead micro-organisms was in widespread use. However, the mechanism by which vaccination protected the individual against a particular infectious agent was not understood, and, as is usual under such circumstances, a number of wild speculations were in circulation. One soon to be discredited theory put forth by Pasteur held that infection by either natural means or vaccination depleted the host of some undefined nutrients needed by the pathogen for survival. Hence on re-infection, the pathogen literally 'died of starvation'!

A more productive advance occurred in 1888 when Pasteur's associates, Emil Roux and Alexandre Yersin, found a *toxin* in the supernatant of cultures of the diphtheria organism that by itself could produce all the symptoms of the disease.[3] This raised intriguing prospects: could the problem of immunity be reduced to a reaction between toxins and antitoxins? Emil von Behring and Shibasaburo Kitasato searched for the latter and quickly discovered a substance in blood serum of animals immunized against this diphtheria toxin that had the power to destroy it and, more importantly, could confer protection against the disease when injected into unvaccinated animals.[4] By the usual crude chemical and solubility tests used at the time the substance could soon be identified as belonging to the class of serum globulin proteins.

Von Behring (1854–1917) was a military surgeon interested in antisepsis and familiar with the disinfectants that could be applied to a wound externally to prevent microbial invasion. He began his scientific career with the idea of finding an analogous 'internal disinfectant' (hoping in particular that external ones might also work internally) but was forced to admit failure after a number of years, concluding that 'before bacteria are killed by a disinfectant or their growth in the organs can be stunted, the infected animal body itself is killed by this same agent'. Von Behring subsequently joined the Institute for Hygiene of the University of Berlin and came under the influence of Robert Koch where his search for 'disinfectants' could be channelled into considerably safer areas, such as using serum from vaccinated animals. But in the interim he was responsible for adding new terms to the language: 'antitoxin', to be replaced by the more neutral term 'antibody' when it was found that only a minority of microbes worked via toxin production, and 'antigen' for the causative agent.

Von Behring's work received much support from Paul Ehrlich (1854–1915), another associate of Koch's in the Institute during this period. Ehrlich demonstrated that vegetable toxins also induced formation of serum antibodies and in his studies of lactating, immunized mice showed for the first time the phenomenon of passive immunity—antibodies could pass from mother to offspring and were just as effective as antibodies produced in the offspring would have been. Von Behring's later studies, including the development of methods for quantitating the amounts of both toxin and antitoxin present in blood serum, led to the foundation of the Royal Prussian Institute for Experimental Therapy in Frankfurt-am-Main in 1899 which was responsible for routine state control of immunotherapeutic agents as well as for research and training in immune therapy.[5]

All this points to molecular mechanisms, but a few years earlier Elie Metchnikoff (1845–1916), a Russian zoologist, had proposed that phagocytosis was the first line of immunological defence.[6] This completely different concept was based on his observations that some leukocytes could engulf and destroy bacteria or other foreign objects. This theory was popular with the followers of Rudolf Virchow's cellular pathology[8] and particularly with Louis Pasteur, who was so impressed that he invited Metchnikoff to his Institute in Paris where the latter spent the next twenty-eight years verifying and defending his phagocytic hypothesis. Metchnikoff's adherents were unimpressed with von Behring's antitoxins, but the humoral immunology cadre was equally unenthusiastic for the idea that leukocytes were responsible for clearing the body of unwanted foreign invaders. The humoral side of the argument had

Fig. 16.1 Paul Ehrlich. Ehrlich and Metchnikoff, joint winners of the 1908 Nobel Prize, advocated opposing theories of the body's defence against invaders. (Source: © The Nobel Foundation.)

significant popular support from practitioners of medicine who, at that time, believed that the accumulation of white cells at the site of an infection was actually the cause of the inflammation and not a defence mechanism.

Von Behring received the first Nobel Prize for Medicine in 1901 and Metchnikoff and Ehrlich in 1908 were jointly rewarded by a Nobel Committee that was clearly 'sitting on the fence' with respect to the humoral vs cellular immunity controversy. But by this time most investigators had opted for the field of circulating antibodies and the protein chemistry

Fig. 16.2 lya Metchnikoff. The other half of the 1908 Nobel Prize.
(Source: © The Nobel Foundation.)

their study entailed—interest in cellular immunity was muted until nearly fifty years later.

The rise of immunochemistry

The medical applications of the new field of immunochemistry were undoubtedly the major driving force for the large number of investigators who immersed themselves in this area of research. The recognition that injection of serum from foreign species could lead to side effects such as serum sickness or, even worse, death from anaphylactic shock, led von Behring to heroic efforts to purify the antibodies from the blood, operating on the assumption that, the purer the antibody fraction, the less deleterious would be its effects. The globulin and albumin fractions were separated and the globulin fraction repeatedly precipitated with ammonium sulphate until a high level of antibody titre was obtained. This semi-purified protein was first called 'paralbumin' and later 'pseudo-globulin'.

Meanwhile, Paul Ehrlich stimulated the scientific community with his theories of antibody–antigen interaction. He recognized that antigen and antibody must fit together like 'lock and key', using the metaphor originated by Emil Fischer for enzyme–substrate interactions. He was remarkably prescient and postulated that the antibody had different domains—one to bind antigen and one responsible for secondary biological effects such as agglutination, precipitation, and lysis of cells by complement fixation. But he also had more controversial ideas, such as his hypothesis that antigen–antibody association was irreversible, arising from the formation of what were essentially covalent bonds.[9] This latter idea stimulated many arguments over subsequent years and we find, as is so often the case under the stimulus of controversy, more fantasy than fact in some of them. The physical chemist Svante Arrhenius, founder of the theory of ionization of electrolytes, wanted to liken the interaction to the neutralization of a weak acid by a weak base, while Karl Landsteiner (1868–1943) and Jules Bordet (1870–1961) were colloid enthusiasts and thought that the specificity could be explained by 'colloidal' adsorption processes,[10,11] even though the latter are normally rather non-specific. In principle, of course, Ehrlich was proved to be right in insisting that a high degree of chemical specificity was essential for the interaction, now known to arise from the arrangement of amino acids in the antibody binding sites and not, as he thought, through the formation of covalent linkages.*

During the first three decades of the twentieth century, experimental work took on an almost wholly molecular orientation. Much effort went

---

* Ehrlich would also prove to be successful at the other extreme of the immunity field with his theory of 'selective' antibody formation, to which we shall return later in the chapter. The germ of this theory was in fact already an integral part of his 1897 paper on quantitation of the diphtheria antitoxin assay.

into studying the course of induction of antigen-specific antibodies, partial purification of these antibodies from blood serum, and study of their chemical interactions with antigens. One of the most prolific investigators during this period was Karl Landsteiner (1868–1943) who published over 300 papers and several editions of his famous book, *The specificity of serological reactions*.[12] He was responsible for demonstrating that any foreign substance linked to a protein could induce an antigenic response; he showed the multiplicity of antigenic determinants on the erythrocyte membrane (the basis for the now familiar blood groups[13]); he investigated the antigenicity of proteolytically cleaved native proteins. But, despite the physiological importance of this work, the important questions about antibodies as distinct molecular entities could not be addressed until a demonstrably pure preparation could be obtained.

It was in this context that the development of new separation techniques for proteins came into play. Arne Tiselius (see Chapter 7) used his newly designed electrophoresis apparatus to separate the proteins of blood serum and demonstrated that the major antibody activity was to be found in the slower moving γ-fraction of the globulins. Continuing studies with his young associate, E. Kabat, who came from America armed with some experience in the field of immunology, confirmed the initial results even more spectacularly.[14] Specific antibody was measured directly on the basis of its function: the electrophoretic pattern for an immune serum was compared with the pattern obtained for the same serum after antibody had been removed by specific precipitation with antigen.

The molecular weight of γ-globulin was shown to be approximately 150 000 and other physico-chemical studies indicated a fairly high degree of molecular asymmetry—it was certainly not the typical 'globular' protein.[15,16] Furthermore, precipitation reactions between antibodies and antigens had convinced most investigators that the antibody must contain at least two binding sites per molecule. An American patent issued in 1936 to I. A. Parventjev demonstrated that limited pepsin treatment was able to reduce the size of the antibody molecule without interfering with its ability to bind antigen, providing the first suggestion of what today we would term a 'multiple domain structure'.[17,18] This feature of the molecular structure was subsequently confirmed and became the single most useful tool in the initial stages of trying to work out the molecular architecture.

The structure of immunoglobulin G

The active component of Tiselius's γ-globulin fraction was renamed 'immunoglobulin G' (IgG for short) as work on its chemistry accelerated. Rodney Porter began work on it in 1946 as a student in the laboratory of Fred Sanger in Cambridge, who was at that point just beginning his momentous project on the amino acid sequence of insulin. Given the size of an antibody molecule, the goal of a complete sequence must have seemed a formidable task, but, if the smaller fragments obtained by limited proteolysis were used as a starting point, it might be possible to glean some important information from partial sequences. An additional problem which no one foresaw at the time was that there would necessarily be sequence heterogeneity in the binding regions of the antibody molecules.*

Twelve years after embarking on this ambitious project, Porter and his colleagues had perfected proteolytic degradation while still retaining binding activity and purified what are now called the Fab (antigen-binding) and Fc (complement-interacting) domains.[19,20] In 1959 Gerald Edelman demonstrated that intact IgG consisted of two different polypeptide chains (termed heavy and light) from which followed Porter's model consisting of four chains, linked by disulphide bonds.[21,22] The complete molecule contained two Fab domains—that is, the two binding sites that antibody–antigen precipitation required. Hydrodynamic studies indicated that the Fab 'fragments' obtained by proteolysis were, unlike the IgG molecule as a whole, compact globular entities. The Fab fragments became useful for studying the antibody–antigen reaction when only a single binding site per molecule was desired in order to simplify the thermodynamic treatment of the association.

How do we account for the asymmetry in the overall molecular shape of IgG, which the individual domains lack? Figure 16.3 shows the now familiar Y-shaped model from the laboratory of one of the present

---

* It is perhaps too obvious to need emphasis here, but some readers may appreciate a reminder that all selection theories of antibody formation require normal immunoglobulins to be horrendous mixtures when examined analytically. The malignancy known as multiple myeloma leads to proliferation of a single type of antibody and was for many years the only source of chemically homogeneous protein.

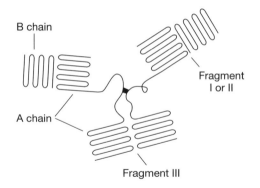

Fig. 16.3 Immunoglobulin shape and domain structure. The model is derived from work in our own laboratory (1965).[23] The present conventional names are 'Fab' for the domains carrying the antibody site (I and II) and 'Fc' in place of III.

authors in 1965, deduced by combining all available experimental data for whole IgG and both domains. Essentially the same model emerged independently in 1967 from elegant electron microscopic investigations by Valentine and Green. The critical functional relevance is the flexible hinge at the centre, which allows the distance between the two binding sites a measure of flexibility.[23,24] A detailed account of all work on active sites during this period is provided in a review by Huston *et al.*[25]

Detailed three-dimensional structures from X-ray diffraction data began to appear in the literature during the early 1970s—the first being an Fab fragment from a patient with multiple myeloma, a neoplastic disease in which the individual produces a large amount of homogeneous antibody thus circumventing the heterogeneity problem inherent in normal IgG.[26] Since that time enormous progress has been made in high-resolution X-ray structures of both fragments and intact antibodies, and many primary amino acid sequences have been published.[27,28]

We have up to this time restricted ourselves to a discussion of the structure of IgG, the most common of the circulating antibodies, but there are actually five classes of antibodies, all with their own unique history of discovery. They are all made up of pairs of heavy and light chains and have much the same overall configuration. One of them, IgM, is a pentamer with five copies of the monomer, which itself consists of the customary two light and two heavy chains.

Table 16.1   Five classes of circulating antibodies

IgA	Found in mucous secretions from the subepithelium of the gastro-intestinal and respiratory tracts; an early antibacterial and antiviral defence
IgD	Located on the surface of developing B cells; no known function
IgE	Found in respiratory and gastro-intestinal mucous and skin secretion Increased in allergic reactions; defence against parasites
IgG	Most prevalent in plasma and extracellular space; now known to represent a secondary immune response
IgM	Found in plasma and extracellular space; primary immune response

Antibody diversity

It was apparent from the days of von Behring and Ehrlich that somewhere in the body there must be cells capable of *synthesizing* antibodies in response to infectious micro-organisms, toxins, or any foreign material. The question was: how could so many different specificities be generated? The earliest theories held that somehow the antigen itself must be incorporated into another molecule thereby conveying an ability to recognize another copy of the same antigen, but this could not account for the fact that far more antibody was produced than antigen injected and that antibody formation could continue for a long time without additional administration of antigen. Much more plausible was Paul Ehrlich's proposal in 1897 that antibodies are naturally produced by cells *prior to exposure to antigen* and reside on the cell surface until shed into the circulation upon reaction with an antigen. Ehrlich further supposed that continuing *specific* antibody production would be expected in order to replenish those molecules that had been removed from the cell surface. This was widely accepted as a good working hypothesis for over twenty years, although it did not answer the question of how to account for the huge number of different proteins that seemed to be required.

But, by the 1930s, chemistry rather than bacteriology had become the prime player in the field of immunology and the chemists reverted to the old idea of antigen transmitting its information to a nascent protein molecule. Breinl and Haurowitz, for example, proposed that the antigen acted as a synthetic template for the assembly of amino acids into the ultimate primary sequence of the antibody,[29] an idea that was entirely consistent with then current speculations about the way proteins in general were synthesized.

Template theories of this kind attained ultimate popularity through the work of Linus Pauling, who had met Karl Landsteiner in New York in 1936 and was immediately intrigued by the question of how such a multiplicity of specificities could be induced in protein molecules. Pauling approached the problem as strictly chemical and in 1940 published a paper suggesting that the antigen served as a physical mould around which a pre-formed (unfolded) polypeptide chain could wrap itself in such a manner as to generate the correct three-dimensional structure.[30] This meant that the antigen had to act as template in the *final stages (folding)* of protein production and therefore implied that antibody could not be produced in a cell unless antigen is actually present, which was contrary to observation, a fact that seems not to have been a deterrent for Pauling. What came to be called the 'instruction theory of antibody formation' claimed adherents among the chemically trained community but was not surprisingly greeted with scepticism by biologists.

The shortcomings of the direct template model were explicitly addressed in 1941 by the Australian virologist, F. Macfarlane Burnet, who suggested an alternative instructional theory in which antigen modifies enzymes used in protein synthesis, so they will continue to make specific antibody even after antigen is gone.[31] Several years later when the probable genetic role of nucleic acids was receiving increased attention, Burnet and Fenner modified the original theory from enzyme modification in the presence of antigen to a direct action on the genome.[32] There is a tendency to refer to these early hypotheses as 'modified template models', but it is important to note that they were a huge improvement in that they allowed continuity in the synthesis of specific antibody long after the antigen itself had gone, whereas the Pauling model required the antigen to perform its template function over and over again, each time an antibody molecule was made.

The first realistic biological selection theory of antibody formation came in 1955 when Niels Jerne (1911–1994) returned to Ehrlich's original postulate of antibodies as cell surface receptors waiting for an antigen to come along to trigger antibody synthesis and subsequent clonal cell proliferation.[33,34] The role of the protein differs dramatically from that envisaged by any of the instruction theories. Globulins differing enormously in configuration and binding specificity are being made all the time—the antigen selects which ones proliferate. Within a few years this idea caught the imagination of numerous other investigators[35–37] and a clonal selection theory which could account for all the known biological facts was rapidly developed.

However, old fallacies are often hard to dispel, and, predictably, some of Pauling's hero-worshippers continued to resist the clonal selection theory well into the 1960s. Templates met their final demise when it was shown unequivocally that fully denatured antibodies regained their specificity on refolding—that is, antigen recognition lay solely in the amino acid sequence.[38] Sequence in turn determines the three-dimensional structure, as is true for all other proteins.[39] In the case of antibodies there is a special feature: overall sequences can be subdivided into so-called 'constant' and 'variable' domains—only the latter show the exceptional variability that is needed to explain functional diversity.

Cellular immunology

Since 1960 cellular immunology has become a burgeoning field of research. The immune process has become seen as a network in which everything depends on everything else and the role of various T and B lymphocytes in the process is gradually being elucidated. The mechanism of antibody-induced cell lysis and agglutination, autoimmunity, diseases of the immune system arising from neoplastic tumours—all these and more are still under active investigation. Details of this ongoing work are beyond the scope of this book even though there are fascinating (and controversial) stories to tell. Suffice it to say that proteins are the controlling factors in these processes, not only as immunoglobulins serving as recognition signals, but also in other specialized roles as well, necessitating a knowledge of their structures and interactions. We make reference to a couple of recent reviews that have a historical component.[40–42]

CHAPTER 17

Colour vision

The peculiarity of light sensation derives not from special characteristics of *light*, but from special activity of the *optic nerve*, which produces only the one kind of sensation, regardless of how its is excited.

H. Helmholtz, 1852[1]

The only satisfactory method of explaining our perception of colours is to suppose that we have in our eyes several different sets of nerves, one set being most affected by one kind of light and another set by a different kind of light.

J. Clerk Maxwell, 1861[2]

Colour vision is the most highly prized of human sensory perceptions. The magic of colours in nature and in art has inspired civilized people from the earliest days. Not surprisingly, no field of animal physiology has attracted more attention from the wise and great, mostly people with no connection to protein science at all—poets and philosophers, physicists and biologists, all have been fascinated by it and have expressed ideas about it. This involvement of celebrities from other fields in a problem that can ultimately be solved only in terms of the photochemistry of proteins forms the principal focus of this chapter. The photochemistry itself turns out at the end to be relatively simple, an example of the familiar effect of chemical environment on the energy levels of organic chromophores, with the molecular fabric of protein rather than a liquid solvent providing the environment—but it took over a hundred years from the original definition of the problem to its molecular solution.

Early definition of the problem was surprisingly correct

We begin with Isaac Newton, the first person to produce the colours of the spectrum with a prism. He recognized that the concept of colour is a construct of the brain, usually, but not necessarily, triggered by light. He asked questions about colour vision in his *Opticks*, first published in 1704, which were intended to be rhetorical, assertions rather than questions. Do not the colours that are 'seen' in the dark by exertion of finger pressure on the eyeball arise from the same motions in the eye as the colours excited by impact of light on the retina?'[3] No one with scientific credentials has disputed the fact. The need to discriminate between the initial stimulus and the events in subsequent processing by the brain became the central doctrine of the great school of neurophysiology that arose in Germany in the nineteenth century under the leadership of Johannes Müller (1801–1858). Hermann Helmholtz, whose views on vision we quoted at the head of this chapter, was one of his disciples.[4,5]

Thomas Young (1773–1829), the brilliant English physicist/physician, addressed himself in 1802 to the mixing of colours, the skill of countless professional painters. It was common knowledge that mixing red and yellow led to the *perception* of orange which the brain recognized as a single colour and it was this phenomenon that Young set out to investigate. His experiments utilized very thin transparent plates, which Newton had shown to reflect different colours, depending on thickness. He concluded that there are three primary colours out of which all others can be created and they must correspond to three distinct receptors in the eye, the signals from which are sent by the optic nerves to the part of the brain that deals with vision. A rare early insight into something that happens in the brain! Maxwell called it 'one of those bold assumptions which sometimes express the result of speculation better than any cautious trains of reasoning'.[7–9]

Colour-blindness

A few years before Young's seminal work, John Dalton (1766–1844), the father of the chemical atom, became in his own person a part of the database for colour vision, because he was colour-blind. Legend has it that Dalton had given his mother a pair of what he saw as conservatively dark blue stockings to wear to Quaker meetings. She was dismayed when she saw them as scarlet! Dalton first attributed this to his mother's senility, but was soon forced to acknowledge that the discrepancy lay in some fault in his own vision. He proceeded to a detailed self-analysis, which formed the topic of a paper read to the Manchester Literary and Philosophical Society in 1794.[10] This is the first report of colour-blindness to which any attention was paid and to this day the French word for colour-blindness is *Daltonisme*.

Dalton was the creator of the first primitive models of molecules; he called them 'compound atoms'. Could he have imagined a protein molecule with its thousands of atoms, or the idea that missing just a single kind of protein could be the cause of his affliction?

Needless to say, colour-blind individuals were a great asset in the game of mixing component colours in order to characterize the physiological receptors, for, typically, such individuals lacked one of the three receptors, leading to a binary rather than a tertiary equation for the summing process.[11] For example, when the great physicist James Clerk Maxwell (1831–1879) became intrigued by colour vision (consciously following in the footsteps of Newton, perhaps) he included data for people who

Fig. 17.1 James Clerk Maxwell at Cambridge in 1855, holding the colour top. (Source: *The scientific letters and papers of James Clerk Maxwell*, Cambridge University Press.[2])

suffered from this affliction and indicated it in the very title of his paper on the subject.[9] Maxwell devised an improved apparatus for colour mixing, called a colour top, and also a clever graphical method for interpreting the results in terms of the three primary colours—the theoretical basis of which his data strongly supported.

Hermann Helmholtz[12,13]

Hermann Helmholtz (1821–1894) was the authoritative voice of physiological optics and his work continued to be regarded as definitive even after he abandoned physiology and became professor of physical chemistry at the University of Berlin. His achievements during the preceding twenty years had been legion. He proposed the law of conservation of energy while still a youth in military service. He was the first person ever to measure the speed of transmission of nerve impulses of any kind. He invented the ophthalmoscope (1850–1851) as a decisive tool for his optical research. He was the author of what until quite modern days was the standard work on optics, his *Handbuch*, published in three parts between 1856 and 1867. Maxwell, in a brief biography written in 1876, called Helmholtz an intellectual giant.*[14]

Helmholtz published his first work on colour vision in 1852, in his *Habilitationsvortrag* at Königsberg University.[16] He repeated Young's work on colour vision (with up-to-date techniques), but at first did not agree with Young's conclusion: he thought five primary colours were required to account for his results. This proved to be an artefact, resulting from insufficient allowance for overlap between the spectral bands assigned to the primary colours.[17] By 1858 Helmholtz had convinced himself that three was the right number and he became Young's chief advocate. The theory of three primary colours became known thereafter as the Young–Helmholtz theory.[18]

We have mentioned Helmholtz's inherited respect for philosophers and his perceived need to acknowledge their work. In regard to Goethe's

---

* When the Cavendish Laboratory was founded in Cambridge in 1871, Maxwell became the first professor and successfully launched the laboratory on to its memorable future. In fact, Helmholtz had been Cambridge's first choice, but he had just accepted the post in Berlin and felt he could not change his mind. Helmholtz's wife is said to have regretted the decision; she would have rather lived in Cambridge than Berlin.[15]

explicit ideas about colour vision, however, Helmholtz felt compelled to be critical. We cannot wish away the need for a sound physical mechanism, he said. 'Wir müssen die Hebel und Stricke [levers and ropes] kennen lernen',[19] even if they upset poetic views of nature. But Helmholtz himself, it must be said, never expressed any curiosity about the possible *molecular* nature of the 'levers and ropes'. The time was too early; the level of sophistication in sensory physiology, thanks to people like Helmholtz himself, was much higher than it was in the still infant science of organic chemistry.

Rods and cones and rhodopsin

The rods and cones of the retina are spectacular anatomical objects, familiar to us all as images from countless textbooks. They were known around 1850 and a definitive account of them appeared in a remarkable paper,[20] published in 1866 by the German anatomist Max Schultze (1825–1876), who has been called 'a consummate master of microscopic technique'.[21] The paper is illustrated by about 100 drawings, showing cones and rods, their connections to strands of the optic nerve and similar detail, for human retinas and for retinas from other mammals, birds, reptiles, and amphibia.

The drawings were supplemented by physiological studies. Schultze unequivocally identified colour perception as localized in cones; rods, designed for vision in the dark, lacked colour discrimination. Schultze knew about and endorsed the Young–Helmholtz theory of three primary colours, but he could not in any of his species find three anatomically distinct kinds of cones.

The first chemical breakthrough with respect to visual receptors came a decade later, with the discovery and isolation of *visual purple*, the photosensitive pigment in the outer rods, which is bleached by light and recovers its colour in the dark. Priority in the discovery belongs to Franz Boll (1877), a student of Max Schultze, who judged the pigment (from frog retina) to be red and called it *Sehrot* ('visual red').[22] A year later Willy Kühne decided that the colour was really purple and gave the pigment its present name, *rhodopsin*.[23,24] Kühne indicated that protein was a likely constituent of the pigment, but did nothing to follow that up.*

---

* The same attitude is evident in Kühne's other name-givings: the terms 'enzyme' and 'myosin' are both due to him, but he contributed little or nothing to associating their functions with protein chemistry.

Thereafter, interest in the molecular basis of vision seems to have waned, perhaps because enzymes appeared about this time and began to monopolize biochemists' attention. Whatever the reason, formal identification of visual purple as a protein of high molecular weight did not come till 60 years later (an unproductive interval of astonishing length) from the work of Selig Hecht at Columbia University in New York, begun in 1937. Ultracentrifugation was one of the methods he used for characterization and this produced an added dividend, demonstrating that the complex absorption spectrum of the 'pigment' (suggesting the possibility of many components) sedimented *in toto* with the protein.[25] By this time the carotenoid prosthetic group had been discovered as the source of the colour by George Wald[26] and Hecht pointed out that this meant the protein had to be a conjugated protein, with the chromophore firmly attached.[27]

The chromophore and the photochemical mechanism

The definitive event for unravelling most of the problems of visual reception was George Wald's elucidation of the chemical nature of rhodopsin, which he began as a postdoctoral fellow in Zürich with the organic chemist Paul Karrer and continued later as professor at Harvard University. He identified the chromophore as a *carotenoid* related to vitamin A and named it retinal.[28] It was found to contain a linear hydrocarbon sequence with five conjugated double bonds, the stable form of which has all its double bonds in the *trans* configuration. The active form, however, as bound to rhodopsin in its dark adapted state, was shown to be the intrinsically unstable 11-*cis*-isomer, stabilized in the native molecule by interactions with the protein fabric.[29] Light absorption transforms this isomer to its stable all-*trans* configuration, and causes it to dissociate from the protein. Huge movements within the protein fabric accompany this and they trigger further events that affect the membrane potential of the retinal membrane. As in all neuronal processes, membrane potential changes create signals for nerve transmission, in this case along the optic nerve to the brain. It is one of the best-studied processes in all of neurophysiology.[30]

It would be yet another 30 years before the cone receptors became available and could be characterized but, once rhodopsin was understood, the probable mechanism for colour discrimination became fairly obvious. Predictably—entirely in accord with Helmholtz's general theories of sensory perception—the mechanism for colour discrimination centres entirely on the light-absorbing chromophores—that is, the very first event in the photochemical chain. In George Wald's paraphrase of the Helmholtz doctrine:

> Action of light on vision is to isomerize the chromophore from 11-*cis* to all-*trans*. Everything else that happens—chemically, physiologically, indeed psychologically—represents 'dark' consequences of the one light reaction.[30]

In line with this doctrine, the cone receptors were found to be similar to rhodopsin, but with different absorption maxima, in the red, green, and blue, respectively.[31,32] The three action spectra, the curves of physiological response as a function of wavelength (which are the same as the absorption spectra), are shown here in Fig. 17.2. The actual chromophore of the three pigments turned out to be the same as in rhodopsin—that is, 11-*cis*-retinal. The colour differences arise from differences in the protein, an example of the effect of the immediate molecular environment on the optical properties of a chromophore, an effect previously

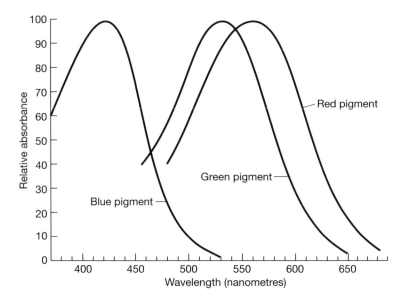

Fig. 17.2 Spectral sensitivity curves for the individual cone receptors. The individual proteins had not been isolated when these spectra were measured: intact cones from cadavers were used. (Based on data of J. Nathans, 1989.[35])

known to protein chemists for many years. It is a property that proteins rarely utilize for any functional purpose,[33] but in this case it provides the essence of the physiological function—a triumph in miniature of the versatility of proteins.

After many more years, the molecular characteristics of the proteins (the four opsins) were finally published. The three cone polypeptides turned out to have about the same length, but the sequences differ significantly from each other and from the rhodopsin polypeptide.[34,35] The data suggest a long history of independent evolution, the evolutionary divergence between red and green being the most recent. Theoretical work on the 'tuning' of the chromophore as a result of the altered amino acid side chains around it continues to the present day.[36]

A final note: the same retinal chromophore is used in another protein for a quite different purpose. Bacteriorhodopsin is a principal protein constituent of the purple membrane of salt-loving bacteria that live in the Dead Sea and elsewhere: its function is to absorb light to create a proton gradient across the membrane, which in turn is used as a source of metabolic energy, beginning with the synthesis of ATP. An outstanding example of nature's economy. And for proteins, yet another instance of versatility!

CHAPTER 18

Muscle contraction[1]

It thus appears probable that the essential condition for extensibility is that actin and myosin should not be linked together. This phenomenon finds a ready explanation in terms of the arrangement of actin and myosin filaments described above, if it is postulated that stretching of the muscle takes place, not by an extension of the filaments, but by a process in which the two sets of filaments slide past each other.

H. E. Huxley, 1953[2]

Introduction

Locomotion is a fundamental property of living organisms. Cells translocate in response to external stimuli, and in higher animals collections of cells can move in synchrony. This unique function is made possible by molecular displacements of protein molecules and sometimes, as is the case in muscle, the scale of motion we observe is millions of times larger than the proteins themselves. Arms and legs and the wings of birds or insects—what we see with our eyes is the aggregate of tiny protein molecular displacements. There is a new dimension involved in asking how it all works. It cannot be solved in terms of the chemistry of a single protein molecule or a single active site for binding or catalysis.

The basic question as to how locomotion is accomplished is, of course, an old one, asked many centuries before atoms and molecules in the modern Daltonian image were even thought of. For example, Thomas Willis and John Mayo were interested in the problem of muscle physiology at Oxford in the mid-1600s; William Croone, one of the founding members of the Royal Society in 1660, went so far as to create an endowed lectureship (still given to this day) to report on 'the nature and property of local motion' and to explain 'the causes and reasons of the phenomena'.[3] Identification of muscle fibres as the source of locomotion in animals also dates back to very early times: Antony van Leeuwenhoek

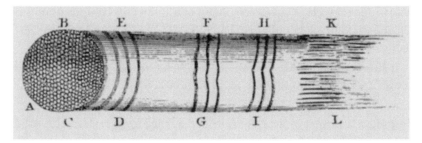

Fig. 18.1 Van Leeuwenhoek's drawing of striated muscle in 1682: his microscopes were capable of magnification up to 270 times. (Source: reprints of Van Leeuwenhoek's drawings by D. M. Needham, 1971.[1])

looked at muscle tissue in his newly developed microscope in 1682 and described not only the fibres, but also the thin longitudinal fibrils of which they are composed.[4] The bundling of the fibrils is evident in his drawings (Fig. 18.1) and so are the transverse striations that give 'striated muscle' its name, striations that might represent 'threads' wound around the fibres, binding the fibrils together—which, of course, is not what they turned out to be. Another famous historical figure, Georges Cuvier, wrote about muscle fibres in 1805, that they are not hollow as nerve fibres are, but instead can be regarded as formed by direct association of the 'essential molecules of the fleshy substance'.[5]

Also going back to early days is the knowledge that muscles, when they contract, can do mechanical work and thus must have an ever renewable energy source available to them. Where does this energy come from? How is it converted into mechanical work? An early classical approach to these questions presided at the birth of the branch of physics known as thermodynamics or energetics. The creative genius was young Hermann Helmholtz (1821–1894), one of the great figures of nineteenth-century science, one of two scientists (the other being Julius Mayer) who can lay claim to being the first proponents of the law of conservation of energy. It was in a paper entitled 'Über die Erhaltung der Kraft', read before the German Physical Society on 23 July 1847, in which the work done in muscular exertion was a cornerstone.[6] Helmholtz found that this work did not exceed the energy of combustion of the foodstuffs consumed in the doing of the work. Helmholtz, of course, was not asking how muscles actually do their work, didn't need to know that muscles were composed of proteins. On the contrary, his evidence may have been stronger for not knowing—even this 'living' engine of still uncertain composition, this engine of completely unknown mechanism, could illustrate the universality of what is arguably the single most important principle in physics.

Justus von Liebig, like Helmholtz an inveterate foe of 'vitalism', cared about chemistry, not physics, and (as usual for him) thought he knew how it was done: the 'albuminous material' found in muscle must itself be used as a sort of intracellular nutrient, burned up in each exertion and then replenished.[7] An absurd idea, but Liebig's forays into physiology were rarely successful. It serves a useful purpose here in that it illustrates a complication for any history of muscle physiology: there are two aspects to the problem, initially conceptually distinct. What molecules within the muscle fibre that we can see in the microscope are responsible for mechanical function, i.e. locomotion? But also, how is energy provided for these molecules—what is the metabolic source? Our purpose in this chapter will be to focus on the problem of locomotion. Energy transduction, basically a part of the more general biochemical field of 'bioenergetics', will receive little attention.

Muscle protein: a century of neglect

Willy Kühne, whom we have met before in relation to enzymes and vision, was the first (in 1859) to isolate a major portion of the muscle protein and to give it a name—'myosin'. The protein was precariously soluble in high salt at 0 °C, and coagulated on warming. Kühne, however, did not believe that this 'albuminous material' directly caused muscle contraction and relaxation; rather, he thought that the sarcoplasm was in some way the seat of the action.[8,9]

There was only sporadic interest in muscle protein in the following decades. Most chemists interested in 'albuminous substances' found it simpler to work with haemoglobin and other easily water-soluble proteins. 'Myosin', in the hands of those who were interested, continued to be prepared more or less à la Kühne, apparently without asking questions as to whether this was a single, homogeneous substance or perhaps a mixture of proteins. This state of affairs persisted into modern times: for example, to Edsall and van Muralt at Harvard, who in 1930 initiated a sophisticated physico-chemical investigation of muscle protein by use of flow birefringence to demonstrate huge molecular asymmetry for Kühne's myosin in solution, the birefringence being generated by molecular orientation along the lines of flow.[10] On the energetic side of the problem, after the importance of ATP became evident,* two Russians, Engelhardt and Lyubimova, made the important discovery in 1939 that ATPase activity resided in myosin itself, again using protein obtained by Kühne's procedure.[11]

Kühne, himself, as we have already noted, had other interests. But in 1898 there was a brief comeback—Kühne delivered the prestigious Croonian Lecture of the Royal Society in London on the subject 'Origin and causation of vital movement'. But he concentrated largely on the topic of innervation: there was no mention of his isolation of myosin 40 years earlier, no mention of proteins at all.

War-time breakthrough in Hungary

At long last, in 1942, there was significant progress on the protein front. An energetic (if overconfident) laboratory director, an inquisitive young man, and the circumstances of war (which helped to leave overconfidence unchallenged) led to the characterization of a second muscle protein, actin, and to preliminary demonstration of its interaction with myosin. The scene was Szeged University in Hungary, the laboratory head was Albert Szent-Györgyi (1893–1986), the responsible author was F. B. Straub (1914–). No revolutionary ideas were involved: just an inquisitive young man meddling with the isolation procedure for myosin. The most significant result was that a 24-hour extraction of muscle tissue yielded 'myosin' solutions of much higher viscosity than a 20-minute extraction, suggesting that a second protein appears during the longer extraction. This was quickly confirmed: Straub called the new protein 'actin'; the low and high viscosity forms of the extract were, respectively, the actin-free myosin and a complex between the two called 'actomyosin'.[15,16]

The initial discovery was quickly followed up. Actin could exist in two forms, a relatively small globular protein (G-actin) and a fibrous form (F-actin) made by extensive association of the former—F-actin was found

---

* The identification of ATP as the critical metabolic intermediate in all of biochemistry is beyond the scope of this chapter, but, since we are stressing the purely mechanical aspects of muscle contraction, it needs to be pointed out that studies of muscle energetics played an important role in the discovery. Classic summaries given close to the time of discovery of ATP are recommended for details.[12,13]

to be the form that combines with myosin to form the viscous acto-myosin. The most spectacular result of all was that actomyosin could be made into long thin threads which could actually be made to *contract* in solution if ATP, K^+ and Mg^{2+} were all present. Myosin alone did not respond under otherwise similar conditions.[15]

Why Hungary, in the midst of a destructive war? A part of the answer is that Hungary (which had become an independent country only in 1918) joined the German alliance at the very beginning of the second war and thereby managed to be left in relative peace for the first few years. Much of the credit, however, must surely go to the leader of the Szeged group, Albert Szent-Györgyi, who had been appointed as professor at Szeged University as part of an attempt by Hungary to breathe new life into its academic scene. He had won a Nobel Prize in 1937 for work on the chemistry of biological oxidation (including discovery of vitamin C), admittedly mostly done while in earlier exile at Cambridge. He was an exuberant figure prone to daring leaps into the unknown; his transition from biochemical pathways to muscle contraction, without previous experience, was entirely in character.

Szent-Györgyi was politically adventurous, too, it should be noted, act-ing as undercover agent for the liberal faction in the Hungarian govern-ment. He went to Istanbul, ostensibly to give a lecture, but actually to contact British and American diplomats to see if anything could be done to save Hungary from Germany's grip. Unfortunately, after his return, the Germans found out what he had done and ordered his arrest—Szent-Györgyi fully expected to be killed. His main concern was for all the laboratory's new data on muscle. He could not face the prospect of their loss, so he sent them to a Swedish colleague (Hugo Theorell) for pub-lication and he himself went into hiding in the Swedish Legation in Budapest. The Swedes also gave him Swedish citizenship and a passport under a false name. His hiding place was soon discovered. He got away (to a new refuge near the Soviet lines), just hours before the Gestapo ran-sacked the Swedish Legation in search of him. This, in summary, is how the aforementioned publication in *Acta Physiologica Scandinavica* arose. The editors, mindful to protect Sweden's neutrality in the war, explained in a footnote that Szent-Györgyi had obtained Swedish citizenship.

Viewed objectively, neither Szent-Györgyi's qualifications nor the Szeged laboratory's facilities were really up to scratch for this kind of work. Were it not for the war, which virtually isolated Hungary from the

Fig. 18.2 Albert Szent-Györgyi, in typically vivacious poses. He was photographed while making the opening remarks at the 1972 Cold Spring Harbor symposium on the mechanism of muscle contraction. (Source: Cold Spring Harbor Laboratory Archives.)

rest of the world, the Szeged group's work might well have been nipped in the bud by legitimate scientific criticism. After the war, a devastating attack did in fact come from Cambridge biochemist Kenneth Bailey. Experimental procedures were shown to be insecure, results often in unresolved conflict with other data, interpretations often in violation of physico-chemical theory. Even the use of the term 'contraction' for the observed dramatic effect of ATP is strictly speaking unjustified, for it could have been a 'shrinking' akin to formation of a fibrin clot.*

But by this time the work had progressed far enough for its benefits to emerge. As Bailey concludes, 'In the last analysis, the hard core of the book, the myosin-actin-ATP interrelationship holds great possibilities for the future understanding of muscle biochemistry and biophysics'. The understatement of the year?[18]

After the war, Hungary became part of the Communist sphere of influence and soon it acquired a regime that rivalled Stalin's in its oppressive government. Scientists fled. Szent-Györgyi and his many co-workers (including his cousin, Andrew Szent-Györgyi) ended up in the United States.[19] Albert quit work on muscle and ventured into purely theoretical

---

* What was observed was in fact not related to physiological contraction at all, but an unrelated chemical phenomenon subsequently named 'superprecipitation'.

projects that proved to be too ambitious for his limited background. But his friends continued in the muscle field and cousin Andrew, in fact, took the next important step in the chemistry of myosin (in 1953): limited proteolysis that split the myosin molecule into structurally and functionally distinct domains—a crucial step on the way to an ultimate understanding of the contraction mechanism, as we shall in due course relate.

We have by this time entered the modern (postwar) era, where the entire arsenal of weapons for molecular protein characterization is thrown into the attack, and where well-equipped laboratories all over the world become participants. In this period, more proteins have been isolated from muscle tissue —notably, tropomyosin by Bailey in 1946,[20] troponin by Ebashi and Kodama in 1963,[21] and, most recently, a giant protein, titin, discovered independently in the United States and Japan.[22] But the actin–myosin structure remains the framework for the mechanism of muscle contraction; the other proteins are present in smaller amounts, providing control mechanisms rather than the basic hardware for contraction and relaxation.

Speculative models for mechanics of contraction

In retrospect, it must surely have been obvious that the muscle fibrils (Fig. 18.1) are the only plausible seat for the cogs and levers of the contractile mechanism. Thomas Huxley, for example, the great apostle for Darwin's theories, seemed to take this function of the fibrils for granted in a popular little book on the biology of the common crayfish, written in 1880.[23] The German physiologist, T. W. Engelmann, was quite certain about it in the 1890s: 'The fibrils are the seat of the shortening power,' he said, and he knew they were 'albuminous' (i.e. proteins) and that 'they always run parallel to each other throughout the length of the fibres'.[24] But there were others who didn't think so—some did not involve the protein directly at all, but thought in terms of movement of the cell protoplasm as a whole, likening it to an amoeba engulfing its food.

Even among those who did involve the fibrils, specific suggestions as to how they worked tended to be pedestrian, hardly worthy of anyone's attention. Colloid chemistry was fashionable in some quarters and produced typical colloid-type speculations, conceptually distant from any involvement with macromolecular structure: changes in surface tension were invoked, for example, or fibrils shrinking and swelling as a result of dehydration or imbibition of water. After ionic charges and the effect of pH upon them were understood to be important in all of protein chemistry, H. H. Weber and K. H. Meyer dutifully proposed electrostatic forces as responsible for fibril contraction and relaxation (ionized groups of like charge repelling, unlike charges attracting) with details extremely vague and no directly related experimental data.[25,26]

The first serious application of the electron microscope to skeletal muscle was reported about this time. Both myofibrils and isolated myosin filaments (from frog and rabbit) were examined, but no enumeration or identification of participating molecules was involved. It was noted that equating the observed myosin filament with the myosin *molecule* leads to a molecular weight of 36 million, but the question of whether this might be the actually functioning molecule was not pursued.[27]

Before the discovery of actin, of course, no realistic model was likely. But even after actin, it took about a decade before the new information (and the new outlook that went with it) blossomed into a credible theory. Meanwhile a single contractile element, 'actomyosin', remained an attractive basis for speculative models. Albert Szent-Györgyi actually gave some thought to the matter and provided a conceivable but unlikely reason as to why two separate proteins might be involved. In most other models, the mechanical core might just as well have been a single protein. William Astbury, famous for his pioneering fibre X-ray diffraction studies of muscle and other materials, actually said (in his 1945 Croonian lecture): 'I should like to suggest that it is not really critical whether what has been called myosin is a genuine individual or not.'[28]

In 1951, Pauling and Corey's α-helix and β-sheet structures entered the scene (see Chapter 11), more fodder for the speculative appetite. Pauling and Corey themselves, elation for their structures outweighing common sense, explicitly suggested a β-sheet → α-helix transition (for myosin alone; actin is not mentioned) as the basis for contraction.[29] Later on, retracting this but still thinking in terms of a single fibre, they suggested a seven-stranded coiled rope, containing both actin and myosin, as model for a new mechanism.[30] And about the same time there was a preposterous new proposal from Astbury, when he was convinced that it did after all matter whether one is dealing with one protein or two.[31]

Most of these and some other models are discussed in a comprehensive review by H. H. Weber and H. Portzehl, published in 1952. Their general (and pessimistic) conclusion is worth citing verbatim: 'Neither the study of the individual purified proteins of the myofibril, nor the brilliant investigations on the fine structure of the fibril, has led to any well-founded theory as to the nature and mechanism of the structural changes which take place on the contractile particles.'[32]

The sliding filament model

As it happened, the correct answer, the sliding filament model, was just around the corner, based on better defined actual pictures, not on hypothetical images. The astonishing fact was that there is no cycle of shortening and elongation of individual molecules at all—neither myosin alone nor actomyosin. Even the filaments visible in the microscope remain unchanged in length. What we have instead is molecules of fixed length sliding past each other, producing much overlap in contraction and less in relaxation. All the earlier fanciful speculations could be assigned to the rubbish heap. There are instances in the history of science where a particularly brilliant hypothesis had predictive value, invoking some principle that was subsequently confirmed. There is no evidence for such here.

The credit for the new model is assigned to two teams, working independently, both based in Britain.[33–35] Hugh Huxley and Jean Hanson did their collaborative work in America (home of the RCA electron microscope), but both were actually on fellowship leave from the Cavendish Physics Laboratory in Cambridge and from King's College, London, respectively. Andrew Huxley (a grandson of Thomas Huxley, but no relation to Hugh) and his German student R. Niedergerke were in the Department of Physiology at Cambridge. The difference in background, physics in one case (with no previous exposure to biology) and physiology in the other, shows up in the style of presentation, but the conclusion is the same.

The genesis of the model can be summarized as follows:

1. We start with the knowledge that myosin and actin are the principal contractile proteins. The first crucial new information came in 1953 with the discovery of two kinds of filaments, termed thick and thin. The thick filament was chemically identified as mainly myosin, the thin filament mainly actin. Actin and myosin, then, are separate morphological entities; no stable 'actomyosin' molecule plays a part in the contraction/relaxation cycle.

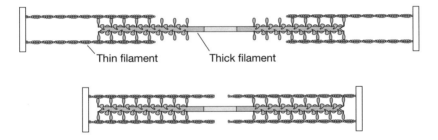

Thin filament   Thick filament

Fig. 18.3 The sliding filament model, based on the 1954 proposal by Huxley and Hanson.[34] [The figure is adapted from the diagram in L. Stryer's *Textbook of Biochemistry* (W. H. Freeman, San Francisco, several editions, 1975 *et seq.*)]

2. Figure 18.3 relates the familiar striations of the muscle fibril to these filaments. The only change on contraction or extension is that the length of the denser part of the A-band (where both filaments overlap) increases at the expense of the less dense part (myosin alone). These alterations in overlap alone account for the shrinking of the length of a sarcomere: actin and myosin filaments don't change length during the process. To quote Hugh Huxley directly: 'Stretching of the muscle takes place, not by an extension of the filaments, but by a process in which the two sets of filaments slide past each other.'[2]

3. Cross-bridges between thin and thick filaments were inferred from X-ray data about the same time and subsequently confirmed by electron microscopy. The cross-bridges belong to myosin: they are visible in reconstituted thick filaments (myosin alone) and absent in actin.

4. The cross-links are asymmetrical; they can tilt along the myosin/actin interface in one direction only! In what is surely a masterpiece of functional design there is a matching alternation in the organization of the filament structure. Myosin molecules point in opposite directions in the two halves of a sarcomere; they pull the actin in opposite directions. As shown in Fig. 18.3, both ends of every sarcomere move towards the centre—we have concerted contraction.

5. In the years since the original model, elucidation of control mechanisms have followed apace, involving additional proteins (for example, tropomyosin and troponin, both mentioned earlier)—tropomyosin turns out to be a remarkably thin long rod, whereas troponin is

essentially a globular protein and contains three distinct subunits.[36] As recently as 1996 the functional role of the giant muscle protein, titin, was identified.[37] It serves as a molecular scaffold for the heavy filament, keeping the myosin molecules in line. It is also an elastic protein, readily undergoing transformation between compact and unfolded states; in the unfolded state it limits extensibility of the fibril—this far and no further, it dictates, no matter how hard one pulls on the muscle itself.[38]

The details of the overall model are described in all modern textbooks and many authoritative reviews.

Heads, tails, and hinges

Identification of the proteins that form the contractile machinery, even with a picture of how they do it on a microscopic scale, is not quite enough in a book devoted to proteins *per se*—a more direct relationship between function and molecular structure is necessary. It came from the dissection of the myosin molecule by limited proteolysis, initiated by Andrew Szent-Györgyi, one of the Hungarian post-war migrants to the United States.

But before turning to this subject, a purely structural detail concerning the whole molecule proved to be historically important: the universal structural element of all proteins, the Pauling–Corey α-helix, does not exist in its pure form in myosin. Although X-ray data for the keratin group of proteins, which includes myosin, played a major role in providing the experimental basis for the α-helix, one prominent reflection in the diffraction pattern does not fit in, as Max Perutz and his colleague Francis Crick at Cambridge were the first to point out.[39] Crick went on to propose that the α-helix can be appreciably deformed, to improve the packing of amino acid side-chain 'knobs' when adjacent helices associate with each other—leading to what has become known as a 'coiled coil', and that this can account for the previously anomalous X-ray reflection. Crick singled out tropomyosin as an especially likely example.[40,41]

With this concept in mind, the picture that emerges from proteolysis is that the myosin molecule is in fact a dimer of two polypeptide chains, which for most of its *length* (but only a fraction of its *mass*) consists of tightly intertwined coiled-coil α-helices. There are two segments of this

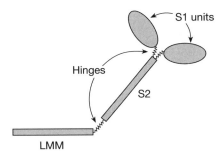

Fig. 18.4 Schematic model of the dimeric myosin molecule. The model is deduced from proteolytic cleavage products. The 'light chains' associated with myosin, but not covalently linked, have been omitted. [The drawing is adapted from a diagram in L. Stryer's *Textbook of Biochemistry* (W.H. Freeman, San Francisco, several editions, 1975 *et seq.*)]

coiled-coil structure, separated by a hinge, as shown in Fig. 18.4. These enormously long segments secure the length of the myosin molecule in the thick filament and have no other chemical function. It is the third domain of the molecule, globular not fibrous, containing most of the mass, that is the chemical end, carrying both ATP- and actin-binding sites. The two heads are not attached to each other in the intact molecule and they emerge as separate entities after proteolysis—the helical segments are cleaved at the hinges, but the dimeric coiled-coil remains intact.

In terms of this model, the cross-links between myosin and actin contain the myosin heads plus the S_2 rod. The power stroke of the contractile mechanism occurs with the cross-link in place: leverage comes from a tilting of the heads by use of the two hinges at the boundaries on either side of the S_2 rod. Quite literally, the myosin filament pulls the actin fibril along itself for increased overlap. Subsequent ATP binding breaks the actin/myosin cross-link, the original untilted molecular form is restored, and a new cross-link, considerably displaced, is formed after ATP hydrolysis. (Displacements in the range of 50 to 100 Å have been reported.) The details of energy transduction—just how the free energy from ATP hydrolysis is fed into the system—is still a matter for debate.

Tropomyosin also turns out to be a long thin coiled-coil double helix, as predicted by Crick.[41] It has been shown to wind along the actin filament; in the resting state of a muscle it blocks the sites to which myosin is attached in the active contraction cycle. The overall model and mechanism present a marvel of engineering ingenuity.

The 1922 Nobel Prize

What we have described in this chapter is the evolution of a picture of what is probably the most complicated function that any protein is ever asked to perform. More than that, it is not just a thing of beauty in itself, but also the solution to one of the oldest and most important problems in all of animal physiology. Yet no Nobel Prizes have been awarded for any of it. Andrew Huxley shared in the Nobel Prize in 1963 for his work on the mechanism of nerve conduction. Albert Szent-Györgyi won a Prize in 1937 for metabolic work related to vitamin C. But no awards for myosin and actin and how they work.

This becomes historically noteworthy when we consider that a Nobel award in the field of muscle *physiology* was in fact made in 1922, shared by A. V. Hill, from University College in London, and Otto Meyerhof, from Kiel in Germany. How did their work fit in with the contractile mechanism that developed subsequently? This question is particularly pertinent in the case of Hill, because he wrote books and chaired discussion meetings that indicated a direct involvement with the central problem of contraction mechanism, whereas Meyerhof was more biochemically oriented. As we are told in the official Nobel award address, Hill's share of the prize was given in recognition of his analysis 'by means of an extremely elegant thermoelectrical method, [of] the time relations of the heat production of the muscle'—a set of intricate measurements that were indeed elegant. But in retrospect they contributed not at all to an understanding of the means by which proteins are able to provide the function of locomotion.

It is true, of course, that, as a physiologist, Hill could have been in the vanguard of muscle research without investigating proteins. He could have been tackling the energetic side of the problem instead, searching for the chemical pathway by which oxidation of sugars or other energy sources could lead to utilization for the work that muscles do. This was indeed what Hill's co-awardee, Meyerhof, was doing, and it led him a few years later to be in on the ground floor of the discovery of ATP as the true key to muscle energetics. But Hill ignored the chemistry of bioenergetics just as much as he ignored muscle proteins. He left that to the bio-chemists, who were at the time off on what proved to be a false trail. (The involvement of organic phosphates was still in the future.[42])

The effect on Hill of the discovery of ATP can be gauged from a 1932 paper entitled 'The revolution in muscle physiology'. Hill was clearly devastated. All of his precise work on heat production in whole muscle had at one stroke shrunk to insignificance. Hill protested that 'all that has changed is a picture of the way in which energy and chemical changes are related', but it was of course more than that—his whole conception of research on muscle energetics had been overturned: it was not physics, not the calorimeter, but organic chemistry that was the essential tool for elucidating energy metabolism. Later on, it would be protein chemistry that would provide insight into conversion of energy into work.[43]

Interestingly, there is no indication that Hill ever changed his course. On the contrary, he quickly recovered his self-esteem and went on for forty more years in the same vein as before, apparently still convinced about the merits of a non-molecular approach—and continuing to gather awards and honorary degrees. He wrote papers on 'negative delayed heat production in stimulated muscle' as late as in 1961, by which time one would think that any possible relevance of heat produc-tion (non-productive for mechanical work) to the central phenomenon would be remote. Hill's definitive retrospective gesture is a 1965 book, which includes a 130-page annotated bibliography. It makes sad reading and offers no explanations for his failure to come to grips with reality.[45]

Concluding comment

Our history has been confined to the striated muscles involved in animal locomotion. But smooth muscles that contract the walls of blood vessels, intestines, etc., work by an almost identical mechanism. In fact, almost all motility, from individual cells to bacteria and upwards, involves actin or myosin or proteins closely resembling them. This is an active field of current investigation—worth a journey for anyone desiring to follow intellectual progress from simple beginnings (chemical identification of proteins) to elucidation of functional design.

CHAPTER 19

Cell membranes

Enough has been advanced here to make it extremely probable that the inorganic composition of the blood plasma of vertebrates is an heirloom of life in the primeval ocean.

A. B. Macallum, 1910[1]

As is well known potassium is of the soil and not the sea; it is of the cell but not the sap.

W. O. Fenn, 1940[2]

Proteins from cell membranes were virtually unknown before 1970. Today they are at the forefront, the cutting edge of the investigation of physiological function. This heart of the subject is part of the 'present' of protein science, history in the making, so to speak. As such, it is not really within the scope of our book, but the beginnings and its links to the past are relevant—was there really no interest whatever before 1970?

The latest textbooks must be consulted for guidance to the current scene. It is not appropriate for us to recommend any particular reference: we have found the introductory book on protein structures by Carl Branden and John Tooze a rewarding place to start. It includes illustrations in the modern style, using different colours to define structural details, and stylized representations of segments of α-helix, strands of β-structure, and the like, which tend to bring both proteins and membranes to life.[3] Our brief comments here will, we hope, help the reader to see the relevance of the past to all this modernity.

The necessity for membrane proteins

In 1881, when secreted enzymes were already well known, but the inside of a cell was still the mysterious material called 'protoplasm', Felix Hoppe-Seyler conjectured that all intracellular chemical reactions were likely to be catalysed by the same type of enzymes as the secreted ones.[4] It is said that someone jokingly responded that this implied that a cell might be nothing but a bag of enzymes, a notion that would have been reinforced some years later by the Buchners' discovery that a mixture of active enzymes poured out of cells when they were lysed by relatively gentle methods.[5] This early use of the term 'bag of enzymes' may be apocryphal, but there is no question that many biochemists treated the cell as such for many decades to come. Their comprehensive atlas of metabolic pathways, giving the biochemical routes by which every known bio-organic compound was made and then converted into something else, was created virtually entirely on the basis of studies of soluble protein enzymes. The biochemists knew about the cell membrane, of course—an essential barrier, needed to keep the cell contents within, while allowing water to pass freely in and out to maintain osmotic equilibrium. But in relation to the metabolic reactions they were studying, the membrane could have been just a container, like the flask they used in the laboratory.

Most biochemists, if challenged, would probably have acknowledged that the idea of a completely inert membrane was rather absurd, but they were understandably too preoccupied with the demands of what they were doing at the time to worry about it. Had they been concerned, they might not have discovered the work of the English physiologist, Ernest Overton (1865–1933), who, in 1899, had spelled out very lucidly the physiological need for active entities within the membrane—for his work was not appreciated: he has been described as 'one of those scientists whose stature is more obvious after their death than it was during their lifetime'.[6]

Overton knew, on the basis of well-conceived and carefully executed experimental work, that cell membranes were made of lipids and that their composition would make them virtually impermeable to polar, water-soluble molecules, because such molecules could not dissolve in the lipid, which they would need to do in order to pass through. Composition alone thus provided a simple explanation for the most vital function of a membrane—enclosure and confinement of the mostly water-soluble cell contents, including inorganic ions as well as the polar organic molecules involved in metabolism.

But Overton had great intuitive ability to add to his skill as an experimentalist. The polar molecules which the membrane prevented from escaping must at one point have entered the cell from the outside and must be replenished as they were used up. That cannot be imagined as within the scope of membrane lipids, for lack of solubility was a passive property without directional preference. Even more important than direction of flow was the fact that inward transport often needs to be 'uphill', from low concentration to high concentration, whereas passive permeation, even if it could somehow be induced, could only go spontaneously from high concentration to low. Metabolic energy is required to go 'uphill' and Overton concluded that the membrane must therefore contain, in addition to its lipids, specially designed engines that not only provide traversing pathways, but must often provide a source of readily available energy as well.

Only proteins can meet these requirements, which Overton could not possibly have known, but which should have been obvious fifty years later. It apparently was not all that obvious and there was actually some resistance to the idea of proteins in membranes from electron microscopists, for the same sort of simplistic reason that led logical positivists around 1900 to reject the idea of the physical existence of atoms by asking 'who has ever seen one?'. Electron microscopists at first had difficulty 'seeing' proteins because membrane proteins, even if firmly anchored to the membrane, tend to move around in the plane of the membrane, making them elusive until special techniques such as freeze fracture were developed.

Chemical principles: cytochrome b_5

Spectacular advances into physiological function began to be made around 1960, such as measurement of ion movements across membranes and the related generation of trans-membrane electrical potentials in the conduction of nervous impulses. As a result, huge numbers of chemists and biochemists become involved in membrane research, but it is not possible to identify a single laboratory or a single discovery as a trigger that opened the flood gates to a true comprehension of how it all worked. Certainly chemical interactions between proteins and membranes could not have been understood without a working knowledge of the hydrophobic factor. While it is true that Irving Langmuir was already fully cognizant of this principle as early as 1917 and even made an effort to enlighten protein chemists about it in 1940, the principle did not penetrate generally into biochemical thought until 1959, when it did so through a perceptive review by Walter Kauzmann, as we have recounted in detail in

Chapter 13. It is likely that many minds turned simultaneously towards membranes at this point. In terms of the total picture, a much-publicized synthesis in 1972 by S. J. Singer and G. L. Nicolson undoubtedly contributed to widespread familiarity with membrane structure, but their work was largely derivative with few original insights—its essential virtue was that it appeared at the right time, just when many people needed a model of some sort to act as a framework.[7]

An early contribution to an understanding of how individual proteins fit into the general scheme can be ascribed to the protein cytochrome b_5, a component of the chain of oxidative proteins involved in detoxification in the liver, a protein that could not generate the sense of excitement that trans-membrane ion channels, for example, would produce a few years later. Cytochrome b_5 was a well-known protein of low molecular weight (11 000), and had even been crystallized, but it had previously always been isolated by some kind of hydrolytic cleavage, suggesting that it might be a part of something larger. This proved to be the case and in 1971 Spatz and Strittmatter at the University of Connecticut were able to use detergents to isolate the parent molecule without degradation. This molecule contained an additional sequence of 40 amino acids, with a predominance of hydrophobic side chains.[8] The meaning was obvious: cytochrome b_5 in its native state must have been anchored to the hydrophobic interior of a membrane, from which it could be liberated intact only by use of detergents, the micelles of which could to some extent mimic the membrane environment. The molecular picture of hydrophilic and hydrophobic protein domains matching their counterparts in the membrane is shown in Fig. 19.1.

About this time we, the authors of this history, became ourselves involved in the ongoing research by elucidating the principles of micellization for different classes of detergent and the principles of the interaction between micelles and membrane proteins.[9] Many of the same insights were obtained independently by two investigators from Finland, Ari Helenius and Kai Simons.[10]

The principle illustrated by cytochrome b_5 was soon found to be generally applicable. The hydrophobic/hydrophilic dichotomy that guides folding of globular proteins was also seen to be the responsible force here. Most protein molecules derived from membranes have been found to contain hydrophobic domains, specifically designed to exist within the hydrophobic lipid interior of the membrane, anchoring the protein there

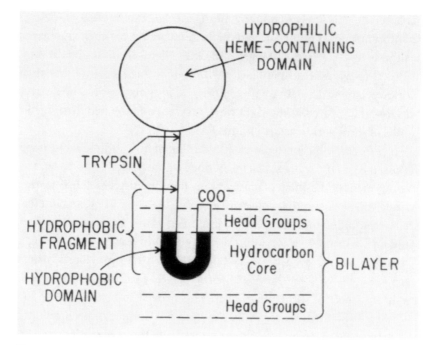

Fig. 19.1 Domain structure of cytochrome b_5. The protein's domains match corresponding regions of a phospholipid bilayer membrane. (Source: from our own laboratory.[8])

because the hydrophobic bit is repelled by an aqueous environment. Domain organization may, of course, be much more complex than in cytochrome b_5. There may be hydrophilic domains on both sides of the membrane; hydrophobic domains can be discontinuous—such as a set of helices running through the membrane and joined outside the membrane by short connecting hydrophilic links, only a few amino acids in length. In the case of cytochrome b_5 the interaction between protein and membrane was just the simplest possible: the protein functioned entirely in the intracellular aqueous medium and the interaction had the sole purpose of an anchor, to hold the protein in place.

SDS gel electrophoresis

The most influential event in practical terms for the growth of the science of membrane proteins was the invention of SDS (sodium dodecyl-sulphate) gel electrophoresis. That is what made membrane proteins 'visible', not quite in the way electron microscopists would have liked, but in a way that chemists could fully understand.

In general terms, the technique of gel electrophoresis is a method by which proteins or polypeptide chains in a mixture are separated and displayed. When a cell is lysed and its contents are examined in this way, only the already well-known soluble constituents appear. When it was learned that detergents can dissolve membranes, they too could be subjected to the same procedure. The detergent SDS was particularly useful because it affected soluble and membrane proteins alike, reducing both to individual polypeptide chains.[11–13] Virtually overnight, new proteins by the hundreds appeared, at first only as sharp bands on a gel, but soon as isolated separate individuals and the object of intense investigation by every conceivable tool.

A remarkable result was that quite different gel patterns were obtained from different cells. Membranes that perform different functions have different proteins! Figure 19.2 is an example, showing the multiplicity of polypeptides obtained from a human red blood cell.*[14]

Physiological function

In the case of cytochrome b_5 the interaction between protein and membrane was the simplest possible: the protein functioned entirely in the intracellular aqueous medium and the interaction had the sole purpose of an anchor, to hold the protein in place. More generally, the physiological functions carried out by membrane proteins are, not surprisingly, along the lines predicted by Ernest Overton. His predictions were not speculative; he was expressing necessity, what a cell needed to do in order to live. Molecular traffic across the membrane is the most obvious function. There are regulated ion channels for 'downhill' movement, often incredibly fast: used, for example, as a mechanism for changing the electrical potential across a membrane in nerve conduction. There are

---

* In Chapter 7 we quoted Arne Tiselius as asking how it could happen that something apparently so commonplace as a separation method should be rewarded by a Nobel Prize. Here we have what is perhaps the most convincing answer: Tiselius's own electrophoretic method, suitably adapted, provided a tool of incredible power, which single-handedly was able to open up entirely new vistas for the investigation of life processes.

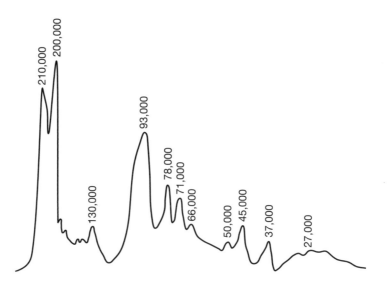

Fig. 19.2 SDS gel electrophoretic pattern for polypeptides from a red blood cell membrane. This was an early result, obtained before optimal conditions for sharp boundaries had been worked out; the boundaries were diffuse and were optically scanned to obtain better resolution. (Source: from our own laboratory.[13])

ion pumps, where translocation is coupled to a source of energy to drive 'active' ('uphill') movement—they are relatively slow, limited in rate by the rate of the chemical reaction that generates the energy, which is commonly ATP hydrolysis. Other membrane proteins use differences between outside and inside ion concentrations as an energy source to drive inward transport of amino acids and other essential organic substances. In plants and bacteria light energy may be used.

A function going beyond Overton's literal predictions is signal transduction, a process mediated by trans-membrane proteins that recognize and specifically bind a signal molecule from the outside, which triggers new chemical events on the inside. The signal is most commonly in the category of a hormone or a neurotransmitter and does not usually itself cross the membrane, exerting its effect via conformational changes in the receptor protein. Hormones and neurotransmitters are usually small molecules; only a fraction (for example, insulin) are proteins.

As we noted at the beginning of this chapter, exploration of these functions is part of the here and now of protein research. Meetings and conferences are devoted to them, new journals appear to present new findings with minimal delay, the internet is used to provide research results even faster than that. The latest textbooks, such as the one to which we have referred,[2] provide essential guidance to the scene, but access to news of current progress—last week's discoveries—is needed to catch the real flavour.

The thoughtful reader will, we hope, see the relevance of the past to all this modernity. The macromolecular nature of proteins (for so many years a controversial matter) is at the heart of everything recognized today. And the concept of hydrophobicity—initially a difficult concept to grasp—has become an infallible guide, essential for understanding every aspect of membrane structure and how proteins fit into it.

Robots par excellence: the sodium/potassium pump

We have suggested that proteins can be considered as somewhat like robots, programmed to perform intricate specialized tasks, with built-in instructions as to when to turn on and off. Cell membranes provide some of the best examples of this concept, often performing tasks of exceptional importance, tasks that are absolutely essential for life as we know it.

Mitochondrial membranes and many bacterial membranes, for example, contain the protein that synthesizes ATP, the all but universal carrier of energy for use in muscles (see Chapter 18) and other situations where a chemical carrier is called for.*[15] The ATP synthase is a complex multi-subunit protein, which we have already mentioned in connection with its subunit composition in Chapter 9. Paul Boyer has called it 'a splendid molecular machine'.[16]

Plants provide another obvious example of a protein complex having a function of central importance for life on earth: the utilization of sunlight in photosynthesis, the ultimate source of all our energy. The central component is the photosynthetic reaction centre, built up of several polypeptide chains plus many pigments. Companions to it are the 'light-harvesting' complexes, which surround the centre and increase the photon-capturing area: not unexpectedly, they contain protein-linked

---

* Mitochondria are often called the 'power plants' of the cells of advanced aerobic organisms. They are thought to have originated from 'infection' of the cells by primitive unicellular organisms like bacteria. In most animals about 90% of the energy derived from glucose oxidation is channelled via mitochondrial ATP synthesis.

chlorophyll to absorb light. Structures of the entire assembly—derived from results for a photosynthetic bacterium—and the manner of assembly of the numerous subunits have been described by Branden and Tooze.[17]

As we noted at the beginning of this chapter, exploration of most membrane functions is part of the here and now of protein research—the structural work on the light-harvesting complex dates from 1991; the detailed structure of Boyer's 'splendid molecular machine' likewise dates from the 1990s. The animal world, however, can provide us with a robotic protein of equally vital importance for the existence of life as we know it, but one that has well-known historical roots, with experimentation and basic ideas going back 100 years. It is, moreover, a considerably simpler protein than the two just mentioned, with only two kinds of subunits and no prosthetic groups—no haems or chlorophyll or other pigments. This is the sodium/potassium pump of animal cells and it is worth a brief account here to give a picture of how delicately precise nature's molecular engineering can be.

To define the pump's task we need to go back to the origin of animal life in ocean waters, a medium which, as everyone knows from personal experience, is rich in common salt (NaCl). In the present context the positively charged Na^+ ion is the important entity: sea water contains high concentrations of Na^+ ions and, conversely, a low concentration of potassium (K^+) ions. The crux of the matter is that within living cells the situation is reversed: K^+ is high, Na^+ low.*

The difference between the sea and living cells has been preserved in an extraordinary way as animals have evolved. All animals are complex organisms, not merely assemblies of cells, but also containing *extracellular fluids*: as in blood, for example, where the working cells are suspended in a fluid plasma. This extracellular fluid has an ionic composition close to that of sea water. In the case of the squid, which was for a long time the favourite experimental animal of the neurophysiologist, the composition

---

* Sodium and potassium exist in sea water, body fluids, and all other aqueous media exclusively in the form of positively charged ions and we shall use the chemical symbols Na^+ and K^+ to designate 'sodium' and 'potassium' throughout this section. The reader should keep in mind that overall neutrality prevails in the long run; a fully detailed description of Na^+ and K^+ movements through a membrane would need to pay some attention to compensating anion movements (for example, Cl^-), which we have not done here.

is almost the same as that of sea water. In more advanced forms of animal life (especially terrestrial ones) the actual values have changed, but a huge intracellular/extracellular difference remains—the cell interior retains the distinction from its environment that it would have had if it had remained in direct contact with the sea.

This preservation of a singular attribute implies, of course, that the sodium/potassium pump must have a vital purpose (evolutionary advantage). Initially this may have been to assure stability of the cell—prevention of water influx and cell lysis—but the overriding factor in the long run was energetic. An ion concentration difference across a cell membrane that is normally almost impermeable to ions is similar to a difference in electrical potential, a source of energy, a force waiting to be unleashed (analogous to an electrical short-circuit) if a conductance channel across the membrane can be opened. In fact, concentration gradients are central to all of bioenergetics, a common feature of the mechanism of all proteins we have alluded to in this section. For example, the ATP synthase described at the beginning of this section uses a hydrogen ion concentration gradient as the intermediate energy form for the chemical creation of ATP. Following the same principles, the difference in sodium concentration between the interior and the exterior of a cell can be used as a source of energy: it is used, for example, as the driving force in the uptake of important nutrients necessary to the cell, such as glucose and amino acids. The single most important function of the pump is probably in the propagation of nerve impulses, in keeping the neuronal membrane 'charged up'. Our knowledge of the overall mechanism of this process derives from the now classic experiments of Alan Hodgkin and Andrew Huxley, carried out in Cambridge between 1939 and 1953.[19]

The energetic driving force for active transport, for the pumping of Na^+ and K^+ against their concentration gradients is (not unexpectedly) provided by ATP and the absolute necessity for maintaining the difference between intracellular and extracellular media is demonstrated by the fact that the quantity of ATP consumed daily by an adult at rest, just to maintain this difference, corresponds to about one-third of all ATP formed for all purposes combined!

The pump protein is a relatively simple protein, without prosthetic groups, composed of two different polypeptide chains, each in duplicate, to give an overall formula of $\alpha_2\beta_2$. The protein was discovered in 1957 by the Danish biochemist Jens C. Skou,[20] and used to be known as the

Na^+/K^+ ATPase, for Skou was not initially measuring ion transport at all—he was simply measuring ATPase activity, the rate at which ATP is consumed. Skou used a preparation of finely ground nerve membranes for his studies, where the existence of distinct intracellular and extracellular spaces had disappeared, but nevertheless obtained an astonishing result, which was that *both sodium and potassium were absolutely essential* for ATP consumption. The protein itself is so purposefully engineered— ion binding and dissociation tightly coupled to the pump's mechanism of action—that the molecular motions it catalyses remain the same even when no possibilty of physiological advantage remains.

The reaction in an intact cell, as subsequently elucidated, has shown that the number of ions transported in the two directions across the membrane are unequal: 2 K^+ ions enter the cell in each reaction cycle, but 3 Na^+ are extruded. Note that this is not a neutral process: one positive charge is lost to the cell per cycle; the pump cycle is intrinsically *electrogenic*.[21]

A note about the transmission of nerve signals

The generation and transmission of a nerve impulse are much faster than would be possible with the Na^+/K^+ pump alone, and the mechanism of action makes use of proteins, called gated ion channels, which open transiently (on demand) to allow rapid flux of either Na^+ or K^+ across the membrane, in directions dictated by the ion concentrations that are maintained by the pump, i.e. inward for Na^+, outward for K^+. Selectivity between the two ions is determined by channel dimensions.

The need for some sort of ion channel had been recognized as early as 1902 by the German physiologist J. Bernstein, who visualized a nerve impulse as a travelling permeability leak, shuttling between the inside and outside of the neuronal cell membrane.[22] The now classic experiments of Alan Hodgkin and Andrew Huxley, carried out in Cambridge between 1939 and 1953, led to the much more sophisticated picture which is now accepted. An excellent summary, accessible to the general reader, was published by Hodgkin himself in 1964.[23] A more detailed account is given in a popular textbook by Kuffler and Nichols.[24]

The experiments of Hodgkin and Huxley were phenomenological, based on imposition of changes in ionic compositions on the two sides of an axonal membrane and the measurement of the resulting electrochemical fluxes. They were done without the knowledge of any of the participating proteins. As late as 1971, in a revised edition of his book *The conduction of the nervous impulse*, Huxley stated that 'At present we have little idea of the nature of the carriers in the membrane, and no real proof of their existence'. As we have seen, the Na^+/K^+ pump protein was actually known by then, though the three-dimensional structure of the pump protein had not been determined. (And still has not been determined, in spite of the protein's ubiquity and great natural abundance.) Structures for some of the gated channels are now known and they are described and discussed in reference 3 and other textbooks.

PART IV

How are proteins made?

CHAPTER 20

The link to genetics

Protein synthesis is a central problem for the whole of biology, and . . .
it is in all probability closely related to gene action

Francis Crick, 1958[1]

Introduction

In this final section of our book, when we ask how the living organism synthesizes proteins, we are entering a new world in which protein science inevitably loses its proprietary identity and merges with genetics and cell biology. And long before our immediate story begins (which is right after the end of World War II), geneticists had already determined that the hereditary material was located in the cell nucleus, that this nucleus contained both proteins and polynucleic acids, and that the

Fig. 20.1 The link personified: 1962 Nobel Prize ceremony. (Photograph by Pressens Bild AB, Stockholm.)

hereditary units, whatever their actual chemical composition might be, were arranged linearly along the chromosomal thread.[2–5]

By 1946 an entirely different breed of scientist began to dominate the scene: self-confident (even arrogant) and conscious of public support. For by 1946 even budget-minded governments had become convinced of the great practical potential inherent in scientific research and in that same year a bill to create a National Science Foundation was before the United States Congress. In Sweden a National Research Council was established with Arne Tiselius as its chairman, and Britain followed soon after with its Medical Research Council Unit for Molecular Biology with Max Perutz at its head. Biological science—perhaps often seen at the government level as an adjunct of *medical* science—had a high priority and even the popular press began to produce leading articles on the subject. Sputnik and the atom bomb may have triggered the surge in public interest and a slice of the public purse, but biology was more benign.

Protein biosynthesis in particular attracted brilliant people, generated almost annual symposia, international in scope, attended by hundreds of participants. The contrast is huge when we make a comparison with the period just a few years earlier, when Fred Sanger was completing his work on the amino acid sequence of insulin. Sanger worked alone on this problem for more than a decade before it was completed, in a small laboratory with a couple of assistants. There were virtually no competing projects on protein sequencing anywhere else in the world; there would have been no quorum for a 'symposium' on such a narrowly defined topic.[6]

The comparison we have made with Sanger's sequencing of insulin is particularly apt to our present subject matter, because 'amino acid sequence' is at the core of the problem of how proteins are synthesized. By 1950, about the time when Fred Sanger was completing his work, it was known that tens of thousands of distinct proteins must exist, most of them containing much longer polypeptide chains than the mere 21 and 30, respectively, in insulin's two constituent chains. All of them almost certainly (as Sanger cautiously asserted) were single chemical substances with unique composition and sequence. This was the period when the elucidation of all the steps involved in the synthesis and transformation of multitudes of organic molecules in living systems was at its height, each step controlled by an appropriate enzyme. But in the case of proteins we have ten thousand or more products of synthesis, similar in that they were all polypeptides, but otherwise as distinct as any other group of organic molecules. Surely there cannot be a separate enzyme for each one? The problem is without parallel in the multitude of metabolic pathways with which biochemists were familiar.

The solution to this problem is part of the great molecular biology revolution, which, as a whole, is described in all modern biochemistry textbooks and some biology textbooks as well. We shall do no more here than to sketch the highlights that particularly relate to protein science. H. F. Judson's book, *The eighth day of creation*, provides an historical account based on interviews with many of the participants,[7] but needs to be treated with some caution since oral history, based on recollections long after the events, can be unintentionally misleading. Personal accounts by many of the participants exist, those by James Watson and Francis Crick being the most famous.[8,9]

One gene, one enzyme

The famous principle, 'one gene, one enzyme', was coined by the geneticist George Beadle, at Stanford University in California, and created a firm link between genetics and the investigation of protein biosynthesis from the earliest days of the latter.[10–12] In a sense it *reduced* genetics to the more manageable proportions of protein science. Conversely, it brought the tools of genetics, transformation of species and the like, to bear on the question of how protein synthesis was managed.

The experimental basis for Beadle's principle lay in the study of genetic mutations induced by means of radiation. This striking phenomenon was originally discovered and purposefully investigated by H. J. Muller at the University of Texas in the United States and resulted in the award of a Nobel Prize in 1946.[13,14] Radiation converted the occurrence of a random mutation from a rare event to a common one: using the fruitfly *Drosophila*, Muller was able to induce hundreds of mutations and to characterize the biochemical changes they caused. What was found was that a mutation was virtually always seen to be created by a defect in a single biochemical reaction, consistent with the interpretation that a single enzyme had gone missing. The rate of mutation was proportional to the irradiating dose and estimates of target size for the impact of radiation were consistent with estimates of the likely size of genes (fractions of a chromosome). Indirect support came from the study of viruses, especially tobacco mosaic virus (TMV), which was becoming a familiar object, both physically and chemically. TMV was a nucleoprotein, about the imagined size of a gene; after infection of a cell it led to marshalling of the biochemical machinery of the cell for production of the virus's one and only constituent protein. It was often regarded as a plausible model for a regular gene, residing in the cell nucleus.[15,16]

'One gene, one enzyme' does not necessarily mean that one gene always corresponds to a unique polypeptide, but the notion was plausible and comfortable for biochemists to live with. A one-to-one relationship between a gene and a particular sequence of amino acids was probably a common supposition.

Besides directing protein synthesis, a gene must also, of course, be capable of making a copy of itself. For this process of gene duplication, the usual conjecture was a 'template' mechanism.[12] We can observe chromosomes creating identical copies of themselves in mitosis every time cell division occurs and we know that the individual genes must have made true copies of themselves when this happens—mutations are inherited with every cell division and continue to be inherited for generation after generation. How is it done? 'Templates' seemed the most sensible possibility. For example, quoting the famous English biologist J. B. S. Haldane:

> The most widely held view is that the gene somehow acts as a master molecule or template in directing the final configuration of the protein molecule as it is put together from its component parts.

and, more specifically: 'The gene is spread out in a flat layer, and acts as a model, another gene forming on top of it from pre-existing material.'[18] A direct connection existed between ideas of this sort and the process of protein biosynthesis because the genes themselves were thought to be proteins. Beadle himself went so far as to assert that replication is just like protein synthesis, except that it's an exact copy of the gene protein that is being synthesized instead of an enzyme.

Protein templates? Are proteins themselves the carriers of genetic information?

The belief that proteins themselves were the templates for gene duplication was quite general. Genes were known to be nucleoproteins, containing both protein and nucleic acid, but the protein component was considered the probable locus of structure-specifying information. Countless examples can be cited, experts and novices alike supporting this notion, up to the very eve of Watson and Crick's 1953 report of the DNA double helix.

Haldane, for example, held to this doctrine; in the quotation just cited[18] the 'pre-existing material' was most likely thought to consist of amino acids. It was the view of mainstream genetics—for example, George Beadle, already mentioned, and Max Delbrück. The doctrine was uncritically accepted by people who had little direct contact with geneticists or biochemists, such as T. Svedberg, a straight physical chemist (and before that, a colloid chemist), who echoed the doctrine in his introductory lecture to a Royal Society symposium on proteins in 1939.[19] His basis was that viruses propagate in living tissue and, plausibly, 'the multiplication is due to an act of autocatalysis, a sort of enzymatic reaction where the virus molecule acts as its own enzyme'. And genes, which he likened to large individual protein molecules, could do the same.

Linus Pauling unquestionably held the belief that proteins held the key to genetics. In the words of his biographer Thomas Hager, writing in 1995, but doing so in partial justification for Pauling's famously incorrect 1952 model for the structure of DNA:[20]

> Protein had the variety of forms and functions, the subunit variability, the sheer sophistication to account for heredity. DNA by comparison seemed dumb, more like a structural component that helped fold or unfold the chromosomes. Beadle believed it. Pauling believed it. At the beginning of 1952, almost every important worker in genetics believed it.[21]

This was at the height of Pauling's great successful ventures into protein structure and function. The seminal papers on sickle-cell haemoglobin appeared in 1949 and 1950; his structures of the α-helix and the β-sheet were published in 1951. In Pauling's mind, DNA was just another structure to be solved, not explicitly seen by him as the ultimate basis of all heredity.[22]

Solid evidence *against* the protein template concept actually appeared in 1944. O.T. Avery, a scientist at the Rockefeller Institute, published a definitive paper in that year with two of his colleagues, which was based on many years of research, and proved that it had to be DNA that was the active agent in the genetic transformation of a bacterium (*Pneumococcus*) from one type to another. By rigorous analysis, no trace of protein could be found in the purified nuclear material that induced transformation.[20] But the result was not accepted by the most vocal and most prominent of the geneticists, led by Max Delbrück at CalTech and others. There was no perceived error in Avery's work, they just did not want to accept it because they had preconceptions to the contrary. They made the lame excuse that one could not really prove that the presence of tiny amounts of protein in the active DNA had been *absolutely* excluded until the same result was reproduced in other experimental systems. Analytical experiment were always subject to methodological limitations.

What militated against acceptance of a defining role for DNA has been explained by historians.[24,25] The main reason was the fact that DNA has only four kinds of constituent nucleotides, and in the early days the four nucleotides were thought to be in close to equimolar amounts. The logical conclusion (first formally attributable to P. A. Levene of the Rockefeller Institute) was what became known as the 'tetranucleotide hypothesis', in which DNA was seen as a simple polymer, with a basic tetranucleotide building block, repeated over and over again along the chain.[25] This suggests a formula for a structural backbone; it seems to hold no possibility for expressing genetic specificity. The protein component of genes was the much more probable locus of structure-specifying information. It was plausible to the majority that 'protein templates' could still be lurking in Avery's nuclear extracts, combined with the nucleic acid that had been holding them in place in the nucleus.*

Needless to say, there was a conspicuous lack of even mildly convincing ideas of how, in practice, a protein template mechanism might work. Delbrück, greatly influential as a geneticist, but lacking credentials as an organic chemist, actually advanced a detailed chemical model for how an accurate copy might arise by means of a free radical mechanism[26]—but no one paid much attention, deservedly so.

---

* Other evidence favouring nucleic acids that was ignored was the action spectrum determined for the induction of genetic mutations by ultraviolet light. This was invariably identical with the absorption spectrum of nucleic acids, not the typical protein spectrum. However, this finding could be dismissed if one wanted to dismiss it—in a complex molecule light energy could be transferred internally before it was used, from the site of initial absorption to another place. There was, of course, no evidence to suggest this was happening, but remote possibilities are the standard refuge of wishful thinkers.

German theoreticians tried to come to the rescue with a putative basis in quantum mechanics. Quantum mechanical resonance, they suggested, would operate preferentially between identical or nearly identical structures. But quantum mechanical resonance was Pauling's own field of greatest expertise and he could see the flaws in their argument and immediately rejected the proposal in principle.[27]

But this did not lead to Pauling abandoning the idea of a protein template. It led instead to adoption of a new catchword, 'complementarity', to try to give a picture of what might be going on. From now on complementarity was deemed to be the pivotal concept, in place of the making of identical copies, an idea that was reminiscent of Emil Fischer's lock-and-key mechanism for the specificity of enzymatic catalysis. Pauling advocated the concept not only for the genetic template, but also as part of the process by which specific antibodies are *synthesized de novo* and appear in the blood as a result of infection—nascent polypeptides of the antibody protein were seen as somehow ductile or pliant, directed by protein–antigen complementarity to fold to an immunologically specific three-dimensional conformation. But Pauling proposed no actual shapes of lock or key for the gene, or for nascent bits of the new copy. Without such detail, the proposal was all empty talk.

The role of physicists

An interesting phenomenon in the early days of molecular biology (as it came to be called) was that a number of scientists trained as physicists changed their horizons and became biologists instead. It is sometimes conjectured that this arrival of physicists in the midst of biologists created a change in the way biologists reasoned and became a vital element in the meteoric achievements of molecular biology. This is probably an overstatement: the quantitative impact was actually quite small.

A legitimate claim can be made for one person, Max Delbrück (1906–1981). He was a 'real' physicist, in the sense that he had a Ph.D. degree in theoretical physics from the University of Göttingen (conferred in 1930) and pursued an academic career in theoretical physics for several years. His early publications were close to the heart of what was still called 'modern physics' (quantum physics, nuclear transformation, and such) and his associates or friends included Lise Meitner, Fritz London, Max Born, etc. He never came to focus on a single area of the subject, however, and was permanently converted to biology, particularly genetics, via the pathway of high energy radiation and the true genetic mutations it was able to induce.[28] A travelling fellowship enabled Delbrück to visit CalTech in 1937 and Nazism and war in Germany led to his remaining in America after that.

On the basis of his work on mutations, Delbrück was led to speculate that genes must have unusual atomic compositions that would normally give them exceptional stability in the cell, so as to require the high energies that radiation provided in order to produce alterations. (Erwin Schrödinger subsequently escalated this statement into a theory of life itself, but there is no evidence that anyone else was influenced.) Delbrück's reputation as a molecular biologist (with minimal influence from his previous exposure to physics) was made instead by his championship of bacteriophages as the system of choice for the study of genetics and the creation of an informal group of biologists that became known as the 'phage group'. Bacteriophages are viruses that attack bacteria and constitute a tool of unrivalled simplicity for the investigation of genetic mechanisms. Delbrück's leadership in the use of this system and the many contributions it made to genetics were acknowledged by his sharing of a Nobel Prize with Salvador Luria and Alfred Hershey in 1969. However, Delbrück's prize came rather belatedly, long after Watson and Crick's, and that is because he did not in fact contribute—may even have been counterproductive—to the mainstream of the molecular biology success story. Delbrück was one of those who clung to the theory that proteins were the carriers of genetic information[24] and did not recognize the importance of DNA, or contribute to the ultimate understanding of how it worked.

But was not Francis Crick, the prime thinker in the elucidation of the DNA story, also a physicist? The objective answer must be that he was not, never a 'real' physicist the way Delbrück was, at least not in any sense relevant to his career. Francis Crick did originally enter science by way of physics, but his education in physics was by his own account old-fashioned and his thesis research unspeakably dull. He subsequently taught himself quantum mechanics, but never found any use for it in his research. Crick was one of the first true molecular biologists, who constitute a breed apart, uniquely different from physicists, biochemists, etc.[29]

The most celebrated physicist whose name always arises in relation to molecular biology is that of Erwin Schrödinger, one of the dominant players in the great revolution in theoretical physics in the early part of the twentieth century. Specifically, he was the originator of 'wave mechanics', the most interesting form of quantum mechanics from a philosophical point of view because of its statistical framework, 'waves' in place of 'particles' and the like. This work had been rewarded with a Nobel Prize in physics in 1933.[30] He was in Dublin during the war at the invitation of the president of the new Irish republic and gave a series of lectures there with the intriguing title, 'What is life', which he published in book form in 1945.[31] Both Watson and Crick acknowledge that the *very idea* that a man of Schrödinger's background and reputation should attempt to write on such a subject aroused their interest, but they and others found nothing there of lasting value. In the book, Schrödinger took notice of Delbrück's comment on the likely 'unusual composition' of genes (see above) and escalated it into a theory that the laws of physics themselves must be different and unusual: 'From all we have learnt about the structure of living matter, we must be prepared to find it working in a manner that cannot be reduced to the ordinary laws of physics.' This was intellectual drivel, completely inappropriate for the middle of the twentieth century.*

One other genuine physicist, George Gamow (1904–1968), will appear in the following chapter, in relation to a single intervention into the problem of the genetic control of protein synthesis—which happened to be historically important, but lacked any depth and did not signal any future efforts in the direction of biology. He was born in Russia, but was a professor in the United States for most of his active career. He was respected in his profession, a physicist whose thoughts were never far from relativity and cosmology, but he was even better known for his attempts to popularize complex physical concepts and for his activities as a jokester—for example, introducing humour even into the titles of his papers and sometimes including fictitious names in his list of contributing authors.[33]

---

* Max Perutz was asked to write a retrospective review in 1987, which turned out to be highly critical. In his own words: 'I accepted [the request] with the intention of doing honour to Schrödinger's memory. To my disappointment, a close study of his book and of the related literature has shown me that what was true in his book was not original, and most of what was original was known not to be true even when it was written.'[32]

DNA and the double helix

It took almost ten years after Avery's experiment to convince nearly everyone that DNA, not protein, must be the carrier of genetic information—that there could have been no cross-contamination between protein and DNA in the genetic transformation experiments that had been done. The immediate cause of the acceptance was the so-called 'Waring Blender' experiments of Hershey and Chase in 1952.[34] The experiment was similar to Avery's; the analytical accuracy was no better—in fact the controls to exclude protein contamination have been described as 'sloppy', not nearly as stringent as Avery's.[35] What was different was that the transforming system used this time was viral instead of bacterial, the bacteriophage infection system favoured by the famous 'phage group'.

Purely chemical analytical studies of DNA by Erwin Chargaff also made a major contribution. They demonstrated that the constituent nucleotides of DNA were not usually present in equimolar amounts and could have quite different ratios in different species—thus destroying any possibility of imagining the DNA as a dull (repeating) tetranucleotide, which could serve no purpose other than to provide a scaffold to which information-storing proteins might be anchored.[36*]

Most important, of course, is that Watson and Crick's double helix was in the offing. The announcement of the probable double-helical structure of DNA by Watson and Crick in 1953 showed, virtually by mere inspection of the structure, the probable mechanism by which the genetic material inherent in DNA could be copied over and over again. In other words, if we *assume*, as the evidence strongly suggests, that the sequence of bases in DNA defines the gene, then the structure itself—the unique complementary pairing of adenine (A) with thymine (T) and of guanine (G) with cytosine (C)—virtually guarantees that the sequence will be preserved.

This was the defining paper of what came to be called 'molecular biology' and it asserted its defining role in what must be one of the most

---

* Chargaff's data also showed that, although the bases did not occur in equimolar ratios, there was a consistent regularity: the amount of A was always essentially the same as the amount of T and the amount of G remained essentially the same as the amount of C. With some imagination could the principle of complementary hydrogen bonds in the DNA structure have been speculatively deduced before the structure itself was known?

significant single sentences ever written in a scientific paper. The bulk of the paper concentrates on the essentials of the three-dimensional struc-ture *per se*, and its consistency with X-ray data, But the classic sentence, at the end of the account, is the most important one:

> It has not escaped our notice that the specific pairing we have postulated immediately suggests a possible copying mechanism for the genetic material.[37]

Protein templates disappeared overnight. Revelation of the DNA copy-ing mechanism did not immediately tell anyone how proteins are made, but it did lead to the disappearance of a lot of vacuous talk that stood in the way of clear thinking on the subject.

CHAPTER 21

After the double helix: the triplet code

It is only within the past 15 years, however, that insight has been gained into the chemical nature of the genetic material and how its molecular structure can embody coded instructions that can be 'read' by the machinery in the cell responsible for synthesizing protein molecules.

Francis Crick, 1966[1]

The fact that DNA must be the carrier of genetic information had become universally accepted and the double-helical structure of DNA had shown, virtually by mere inspection of the structure, the probable mechanism by which the genetic material can be perpetuated, copied over and over again as cells divide. But these revelations did not immediately tell anyone how proteins are made: the difficult to sustain concept of protein templates disappeared from the scene, of course, but what was to take its place? Several good accounts of how the problem was solved were written in the mid-1960s, while actual events were still fresh in the writers' minds. We make reference here to three that are suitable for the general reader.[1-3]

The sequence hypothesis

Combining the statement that DNA alone must be the self-perpetuating carrier of genetic information with the old 'one gene, one enzyme' principle leads to the at first rather astonishing conclusion that the *sequence of bases* in any section of cellular DNA—perpetuated by the complementarity of the double helix—must uniquely determine the *sequence of amino acids* in a corresponding polypeptide chain. This is Crick's 'sequence hypothesis', as given in an anticipatory paper to an audience of biologists in 1958, not as an original idea of his own, but as a hypothesis that was already 'rather widely held'.[4] What is astonishing about it is that there was not at the time any imaginable direct chemical relationship between DNA and protein that might be used to create a synthetic pathway.

Was Crick even thinking in chemical terms? Consciously or not, he seems to be parting company with traditional chemistry in the very terms that were being used to define the problem—an admission that one is not *primarily* interested (as any good chemist would normally be) in 'balanced' chemical equations, accounting for the fate of every involved atom. The emphasis has switched instead to 'information content'. A new non-chemical language had to be invented just to express the idea; a metaphor based on 'language' itself—molecular information portrayed as analogous to letters of the alphabet and their ability to create meaningful words.*

Numerology pertinent to the genetic code

Numerology was probably a factor in many people's minds right from the start. There are only four different nucleotides in DNA to use for building a sequence. But 20 different amino acids (or possibly even a few more) were accepted at the time as building blocks for proteins. A sequence of two nucleotides, viewed as a 'word', as a piece of information, can produce only $4^2 = 16$ distinct entities, not enough to specify 20 distinct amino acids. On the other hand, a trinucleotide would be more than enough— that is, $4^3 = 64$. That is a huge excess: would nature be so wasteful?

The first explicit use of this numerology surprisingly came from the Russian-born theoretical physicist, George Gamow,[5] professor at George Washington University, one of the original proposers of the 'big bang' theory for the origin of the universe, but probably best known for his attempts to popularize complex physical concepts, such as relativity. He seems to have caught the excitement generated by Watson and Crick's double helix, and within a year came out with a paper, which was naïve (even muddled?), but included this numerological aspect that was an essential part of the overall problem.[6]

Gamow's paper was actually very conventional, the simplest possible extension of a direct template mechanism, with the DNA helix itself as

---

* Crick makes a distinction in the overall genetic process between flow of energy, flow of matter and flow of information. Considering the last as an independently definable attribute of what is ultimately a chemical process is indicative of the originality of Francis Crick and other leading molecular biologists.

the template instead of some hypothetical template protein. Gamow proposed a lock-and-key relationship between various amino acids and 'holes' created by nucleotides when they were organized within the DNA double helix. 'One can speculate that free amino acids from the surrounding medium get caught into the holes . . . and thus unite into the corresponding peptide chains.' He showed that the four nucleotides in the Watson and Crick DNA structure could in fact form 20 different 'holes', which proved to be a purely fortuitous result, unrelated to the mechanism. In physico-chemical terms, Gamow's proposal had a flimsy foundation. He made no attempt to measure the size or shape of the helix's structural cavities, nor did he try to relate them to the shapes of different amino acid side chains; and there were other obvious deficiencies in his scheme which we don't need to mention explicitly. The point to be made is that it was Gamow's paper that galvanized Crick and others into action along the lines of an abstract 'code', the design of a table of equivalents, a dictionary of nucleotide sequences versus the amino acid residues they commanded—with initially no idea of how one proceeded from one to the other. According to Crick, it was an advantage to him that Gamow's idea was 'not cluttered up with a lot of unnecessary chemical details'.

Even in the absence of the clutter of chemical details there were lots of fascinating questions to be answered. Assuming that the unit of information at the DNA level (which came to be called a 'codon') is in fact a triplet of nucleotides, what points in the long DNA sequence signal the start and the end of a particular polypeptide chain? Are there 'commas' between codons? Or, contrariwise, do successive triplets overlap, so that one nucleotide can be part of more than one codon? Is there redundancy in the code, such that more than one triplet codes for a given amino acid? If not, what is the role of the excess implied by the 64 possible triplets?[7]

Meanwhile, intense research was of course going on into the actual biochemistry that might be involved between the genetic code and its ultimate expression in the form of proteins.[8] Crick, in his 1958 paper,[4] had, besides the sequence hypothesis, proposed what he called his 'central dogma', which states that information can pass from nucleic acid to nucleic acid and from nucleic acid to protein, but not from protein to protein or from protein back to nucleic acid. This dogma says that the final step is irreversible and also had the effect of eliminating any possibility of protein templates. Crick acknowledged that his dogma did not have universal support, but there is no doubt that it was believed by most and was an implicit guide for all future productive research.

In any case, from this point onwards, events unfolded with an unparalleled sense of urgency. Now forty years later, it is textbook material, at the very heart of all biochemistry.[9] Good detailed accounts can be found in any modern textbook and we need only sketch the bare outline here. For readers desirous of following the progress at the time it happened, in the words of those who were at the very centre of the work, we strongly recommend, in addition to the more popular accounts to which we referred at the beginning of this chapter,[1–3] the Proceedings of the *Cold Spring Harbor Symposia on Quantitative Biology* for the years 1961, 1963, and 1966.[10–12]*

Cellular RNA

The presence of another type of polynucleic acid in cells had been noted. This substance, RNA, is like DNA in that it is a polymer of four nucleotides, with ribose instead of deoxyribose as the constituent sugar. One of the four RNA bases, uracil, differs from thymine, its DNA counterpart, by lacking one methyl group; the others are the same as in DNA. RNA sequences can thus be described in terms of a four-letter alphabet: A, G, U, and C, analogous to the A, G, T, and C used for DNA.

In the years immediately following the Watson–Crick double helix, two kinds of RNA were established entities. Ribosomal RNA was found in the cell cytoplasm and represented around 80 per cent of the total RNA; it also appeared to be the site of protein synthesis. But how could this be, since the DNA providing the code for proteins is located in the cell nucleus? A second type of RNA was a much smaller soluble molecule that appeared able to bind amino acids and represented approximately 15 per cent of the total RNA. This was later named transfer RNA (tRNA) when specific members of the class, one for each amino acid, were found

---

*The Cold Spring Harbor laboratories on Long Island, just outside the city of New York, were an informal meeting place for molecular biologists for several years, quite apart from being the venue of these symposia. Max Delbrück and others often worked and communicated there, away from their parent institutions, over the summer months. American commentators regard it as a historical shrine, almost the equal of the MRC Laboratory in Cambridge.

and it was speculated that these amino acid/tRNA complexes provided a reservoir of activated amino acids for incorporation into a new poly-peptide chain.[13–16]

However, some additional entity was needed to carry the information from the nuclear DNA to the ribosomes in the cytoplasm. This sub-stance, now called messenger RNA (mRNA), was proposed by Jacob and Monod in 1961 and found experimentally the same year. It was a rare joint project from three countries: Jacob from the Pasteur Institute in Paris, Sidney Brenner from the MRC Laboratory in Cambridge, and Meselson from CalTech.[17,18]

Messenger RNA, then, is the missing link between the nucleus and the actual place of manufacture of protein: a copy of the information found in the DNA gene sequence, the 'transcript', as it were. This is then 'translated' in the cytoplasm by the ribosomal particles and the activated tRNA/ amino acid complexes.

This bare-bones outline, of course, does not do justice to the enormous complexity of the process, for at each step there are dozens of proteins involved—proteins that must guide the correct transcription of DNA to RNA, enzymes that process the mRNA after transcription, other enzymes that function during actual polypeptide synthesis on the ribosomal RNA, and more almost *ad infinitum*. This may be the best illustration of the concept implicit in our title: proteins as robots, each doing its precisely defined task in the scheme.

Perhaps the most amazing result of the scientific efforts of these few years was the finding that the whole apparatus could be removed from a cell and would function outside it,[19] always truly following the course that it was instructed to follow by the message. Cell-free protein synthesis had arrived.

Experimental determination of the code

The ultimate goal for protein chemists must be the actual one-to-one relationship between codons and amino acids. This was accomplished more quickly than one might have expected. As stated in the foreword to the 1966 symposium at Cold Spring Harbor:[12]

> In 1961, the genetic code was known to be made up of three-letter words and the word UUU was found to code for phenylalanine. Five years have passed, and the code is now known, to all intents in its entirety.

The symposium proceedings give a complete account of the decipherment, just at the right moment, when it was effectively all done. All the important participants were there, with candid photographs of many of them.

The breakthrough work came from Marshall Nirenberg and J. Matthai at the National Institutes of Health in Bethesda, MD—not members of the inner circle of molecular biologists, such as Crick, Brenner, Monod, etc.*

Nirenberg and Matthaei discovered that a purified cell-free system of ribosomes plus soluble RNA (plus enzymes, cofactors, etc.) was not enough to induce amino acid incorporation: a template RNA was an absolute requirement. More important, the natural template, messenger RNA, was not essential—synthetic RNA would also trigger *de novo* synthesis of protein. This allows the use of a synthetic homopolymer (containing just a single species of nucleotide) as 'message', in which case the 'protein' obtained should contain only one amino acid. That was the big result of the paper: poly-uridylic acid (poly-U) gave rise to poly-phenylalanine. A note added in proof reported that similarly poly-C specifically mediates incorporation of L-proline into the synthetic product. Assuming a triplet code, two codewords in the dictionary had been established: UUU corresponds to Phe, CCC corresponds to Pro.[20]

This initial success was followed by use of RNA copolymers as templates. Random mixtures of two or three distinct nucleotides were used to form the copolymers, so that several different triplets were generated and more than one product resulted per experiment. Statistical analysis of the results on the basis of copolymer composition aided in the probable identification by the Nirenberg group of RNA codewords for a total of 15 amino acids. Additional codewords came from similar studies by Ochoa and co-workers in New York and from G. Khorana in Wisconsin, who was able to synthesize RNA copolymers with strictly defined base sequences for use as ribosomal triggers.[20–23]

What finally solved the entire problem so quickly was the triplet binding method of P. Leder and M. W. Nirenberg. They found that binding of

---

* The successful experiment with polyU was done in May 1961. There was a Cold Spring Harbor Symposium (see above) shortly thereafter in the summer of 1961. Nirenberg had applied to attend, but was turned down! This shows how much of an outsider he was, not a member of the 'club'.

Fig. 21.1 Marshall Nirenberg. (Source: courtesy of the National Library of Medicine, Bethesda, MD.)

each specific tRNA (with attached amino acid) into the ribosome preceded incorporation into the growing chain and was much easier to measure than waiting for synthesis of polymer or 'protein'. Any trinucleotide incorporated into a ribosome would induce binding of the corresponding tRNA; dinucleotides never worked. This was ultimate proof of the triplet hypothesis, if one felt there was still need for such at this stage, as well as, of course, adding to the 'dictionary', the pairing of the codon (in RNA language) with its corresponding amino acid.[24,25]

The upshot of this and related work was that the code is now known to be universal for all forms of life; '3', the minimal number calculated by

Table 21.1   The RNA code*

1st position	2nd position				3rd position
	U	C	A	G	
U	Phe	Ser	Tyr	Cys	U
	Phe	Ser	Tyr	Cys	C
	Leu	Ser	Stop	Stop	A
	Leu	Ser	Stop	Trp	G
C	Leu	Pro	His	Arg	U
	Leu	Pro	His	Arg	C
	Leu	Pro	Gln	Arg	A
	Leu	Pro	Gln	Arg	G
A	Ile	Thr	Asn	Ser	U
	Ile	Thr	Asn	Ser	C
	Ile	Thr	Lys	Arg	A
	Met	Thr	Lys	Arg	G
G	Val	Ala	Asp	Gly	U
	Val	Ala	Asp	Gly	C
	Val	Ala	Glu	Gly	A
	Val	Ala	Glu	Gly	G

*In DNA, thymine (T) replaces uracil (U)

Gamow, turned out to be the right number, giving 64 possible codons, all of which are used in some way. The code is redundant; several triplets specify the same amino acid, 61 out of the 64 possibilities are actually used in protein synthesis for insertion of amino acids into a growing polypeptide chain. And the three that are 'nonsense' in dictionary terms are retrospectively seen as just as essential as the other 61—they are needed to signal the end of a polypeptide chain, the point where synthesis of a particular polypeptide ceases.

The new alchemy

Prolongation of the physical life – with immortality as the ultimate goal—was the aim of the first phase of alchemy.

O. B. Johnson, 1928[1]

We declared in the introduction to this book that we would not try to see into the future and we have not altered this intent. But the future in protein science is being kept perpetually before us in this millenium year, pushed into prominence by newspapers and popular magazines—abetted, it needs to be said, by quite a number of scientists themselves, in whose pronouncements a touch of modesty might be more becoming. Some extravagant promises are being made, reminiscent of the Middle Ages, when the alchemists tried to point the way to everlasting bliss. At the inorganic level they sought to transmute base metals into gold, in biology they sought the elixir of life, essentially the infinite prolongation of life. Now, many centuries later, the physicists have managed atomic transmutation of a sort (though not from lead to gold) and the belief is rife that we may be on the verge of the elixir of life as well.

The roots of this new alchemy go deep into the past, as much as fifty years, and we think it appropriate to sketch these links to the present, even if only to reassert the old adage about everything we do resting on the shoulders of our predecessors (which applies to crystal gazers as much as to those who stick to the laboratory bench).*

Evolution: proteins and species

At the heart of this new alchemy is the genetic code: the ability to do our thinking and often even to carry out our experiments interchangeably in terms of the amino acids along a polypeptide chain or the triplet codons

---

* See, for example, Gratzer, W. (2000). *The undergrowth of science*, Oxford University Press; especially 'Some mirages of biology', pp. 29–64.

that specify the amino acids in the nuclear DNA. Protein science and genetics have merged, so to speak. And genetics includes within it both the short-term events within an individual's life-time (gene expression) and the long-term changes from generation to generation—in other words, evolution. In this brief chapter, we take up evolution first.

We have known since Sanger's determination of the amino acid sequence of insulin, which was carried out in 1949–1955 (see Chapter 8) and which included insulin from more than one species, that changes in protein sequences must provide the essential chemical basis for generating evolutionary change and most protein scientists were probably ready to accept this principle even earlier as the only reasonable hypothesis. But firm knowledge of the genetic code has had a dramatic effect on the underlying conceptual framework. The meaning of mutation has become clear: for example, change GAA to GUA in the base sequence and valine replaces glutamic acid in the peptide chain; change GAA to AAA and the replacement is now lysine. And the underlying process of translation (involving RNA and ribosomes and a battery of enzymes) is the same for all amino acids and all living species.

Francis Crick predicted in 1958 that amino acid sequences could be exploited for exploration of genetic and evolutionary purposes. In his own words:[2]

> Biologists should realize that before long we shall have a subject which might be called 'protein taxonomy'—the study of the amino acid sequences of the proteins of an organism and the comparison of them between species.

But it needed the dual fluency in genetic and amino acid languages to bring this about, to speculate on how proteins evolved from one another, how new functions might arise, for example, from just a few changes in the sequence of base pairs. We can now imagine a molecular taxonomy that can encompass the whole spectrum of living organisms: families and genera of proteins treated as independent of the species they inhabit. One writer (Cyrus Chothia at Cambridge University) has examined a large volume of DNA sequence data and has suggested a clustering into families of this kind: a limited number of groups within which there are close similarities. He makes the encouraging estimate that the total number of existing protein structural families (common ancestors) may be less than

1000.[3] The new 'structural genomics initiative' that we mentioned in the preface to this book put the number at 10 000.

Starting from a different perspective, we can use proteins to speculate about evolution of species: traditional evolutionary trees based on morphological characteristics can be augmented with molecular versions based on the amino acid sequences of proteins of related species. Early efforts to exploit amino acid sequence determination for this purpose date from the 1960s: one of the first made use of the protein cytochrome *c*, which has only a single short polypeptide chain and has the advantage of being essential for all forms of life, from primitive unicellular organisms to humans.[4] By itself, the effort was not spectacularly successful. Without secure knowledge of how three-dimensional structure relates to sequence and how function relates to structure, it proved to be difficult to distinguish functionally beneficial mutations from those that are 'neutral'. Ability to interpret alterations in primary protein structure, however, improved when sequence data were later supplemented by structural differences obtained by X-ray crystallography, with commensurate increase in the labour that was involved.[5,6]

Now that we can skip back and forth between genes and the proteins for which they code, the comparison becomes more practical—the exact relationship between molecular and morphologcal trees can be established and the ability to specifically alter the genome* opens the way to major advances in our understanding of evolutionary differences. The possibilities for practical applications are widespread and cover such diverse areas as agriculture, animal husbandry, and human disease.

The application of greatest interest to the general public is, of course, the application to human genetic disease. The 'elixir of life' in action, so to speak: news stories headlining identification of genes that may be responsible for Alzheimer's disease and other feared perils.

Genes and human disease

The most famous example, historically, of the link between a single site mutation and disease originated way ahead of its time, as far back as 1949, long before the genetic code and its implications were elucidated. It was a time when even constancy of amino acid compositions for given proteins was still being debated; when the amino acid sequence of

---

* The usual technical term for such changes is 'site-directed mutagenesis'.

insulin was just in the process of being completed. The demonstration of the link came from Harvey Itano, working with Linus Pauling and some of his other associates on the identification of a human molecular defect as the cause of the disease of sickle-cell anaemia.[7]

Red blood cells in this disease change their shape at low levels of oxygenation, from the normal biconcave disc to a sickle-like crescent, and they become rigid. In this state they are fragile and their destruction leads to anaemia. The effect is reversed by oxygenation and (when studied *in vitro*) by carbon monoxide. The obvious implication is that haemoglobin is itself the seat of the disease, the only cellular component subject to reversible change as a result of binding oxygen or carbon monoxide. This was confirmed by use of electrophoresis—the old-fashioned boundary electrophoresis of Tiselius, for the now familiar gel electrophoresis would not become available for another decade.

Electrophoretic mobilities for normal and sickle-cell haemoglobins, measured as a function of pH, were found to be different. The effect was especially striking at pH 6.9, close to the isoelectric point, where normal haemoglobin molecules are anions, whereas the sickling variant molecules were found to be cations and moved in the opposite direction. Quantitatively, the difference in molecular charge needed to account for the observations was estimated as 'about three charges per molecule'.[8]

Much more precise definition of the difference was obtained by Vernon Ingram at Cambridge University, who followed procedures that had been initiated by Fred Sanger for his determination of complete amino acid sequence. Paper chromatography of the products of a complete hydrolysis was the first step, producing dozens of spots on the paper, but it indicated that only one unique spot was affected. Analysis of the peptides corresponding to that spot showed that just a single amino acid was different between the two proteins: a glutamic acid residue had been replaced by one of valine.[9,10]

Haemoglobin is a tetramer, containing two α and two β chains per molecule, but at the time this work was done the amino acid sequences of the individual chains were not yet available; one could not even know from which chain the mutated peptide fragment had originated. Nevertheless, the suggestion was strong that the glutamic acid replacement was the only change that had occurred. Given the polypeptide chain composition, the variation would have to occur twice in each molecule, creating a change of 2 in the net molecular charge, close to what Itano *et al.* had

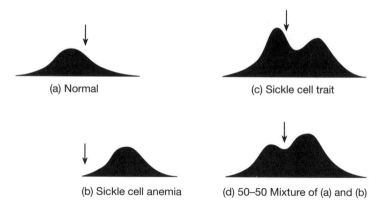

(a) Normal   (c) Sickle cell trait

(b) Sickle cell anemia   (d) 50–50 Mixture of (a) and (b)

Fig. 22.1 The molecular flaw in sickle-cell haemoglobin, defined by electrophoresis in 1949. The conditions used were such that positively charged protein ions would move to the left and negatively charged ones to the right; arrows show the position of the original boundary. (Based on Pauling *et al.*[7])

estimated. The direct cause of sickling and cell fragility is intracellular precipitation of haemoglobin. Quite understandably, the substitution, 2 Glu → 2 Val, will change the charge distribution 'towards one favouring abnormally easy crystallization'.[11]

The overall conclusion, that sickle-cell anaemia is caused by replacement of a single amino acid, presumably by a single point mutation, was of course sensational. This was the first time in any organism that a single point mutation was identified as the cause of a disease and the chemical nature of the mutation even provided a plausible explanation of the symptoms—one electrostatic charge on one amino acid side chain is enough. Other abnormal haemoglobins were subsequently discovered by electrophoresis or other means and given rather unimaginative names (haemoglobins C, D, E, etc.), and most proved to be the product of an allelic gene—that is, a likely candidate for another point mutation. It was also found that human fetal haemoglobin is *normally* different from the adult variety, but in this case many amino acid replacements were found to be involved. (A review by Itano lists all the genetic variants that were known at the early date of 1957.)[12]*

---

* Are the new alchemists generally aware of this remarkably early instance of the precise genetic definition of a human disease? Some may of course consider it of small import: it defines, but does not point the way to what to do about it. The truly dedicated alchemist wants to change the world, not merely to understand it.

The brave new world

Today there is a lengthy catalogue of human disorders attributable to incorrect protein sequences (and hence to incorrect DNA sequences) that might be amenable to genetic manipulation, but despite the optimistic front page headlines, such advances probably lie well in the future. Human physiology is considerably more complex than that of single-cell micro-organisms, the milieu in which molecular biologists have the greatest amount of experimental experience.

But genetic manipulation is not the only way forward. Instead of changing a deficient gene—for example, one that is supposed to make insulin but doesn't—we can supply the missing protein from an outside source. In the case of insulin this has already been done for many decades, extracting this protein from other animal sources, and knowing from experimental tests that the foreign hormone works as well as the native one. In this case, however, we can now go one better: *we can grow the human form itself*, using genetically altered micro-organisms that (before they were altered) had no need for insulin and no gene at all devoted to its production. A major biotechnology industry has grown up around the production of such specific proteins, primarily for pharmaceutical use.

One can go even further afield, beyond disease, to normal cellular differentiation and development and to cell ageing and death. Major advances in the recent past have identified scores of proteins involved in these processes. Will it be possible to control them by some outside intervention? Research into such possibilities is already under way in the form of treating undifferentiated cells growing in culture, with the aim of inspiring them to turn themselves into specialized cells such as are found in nerves, muscles, etc.

One can even think in terms of designing proteins *de novo*, proteins *à la carte*, that have some particular beneficial characteristic, such as an improved binding site for an enzyme, and this protein can then be synthesized outside the cell. Here the actual sequences of whole individual polypeptide chains have drifted into the background. Instead the focus is on partial sequences that have been identified as tending to fold to the desired active site—computer programs pick the partial sequence that has the highest probability of folding to the desired active site, which may not be closely related to any real protein at all. In other words the protein as a whole has become just a scaffold for the active site.[13–15]

A final word

For those of us who entered protein science in the middle of the twentieth century and who lived through and/or participated in many of the dramatic events (advances, retrogressions, and controversies) that we have described in this book, the world of molecular genetics has seemed to be not merely a 'brave' new world, but a 'strange' new world—sometimes smacking of sorcery. We remember the dogged (almost plodding) hard work that was necessary to wrench any reliable data from our work: trustworthy molecular weights, amino acid sequences, the first three-dimensional structures, molecular mechanisms for physiological function—even the earliest ventures into molecular disease, the elucidation of the difference between normal and sickle-cell haemoglobin, falls into this category.

Can we trust the new alchemy, where it is all done at a dizzying pace by indirect means? Amino acid sequences are deduced instantaneously from gene sequences and three-dimensional structures are obtained by analogy—a computer is used to scan a database, to search for similar sequences or segments of sequences in proteins of known structure, and, abracadabra, it spews out the desired result. It seemed for a while that proteins *per se* might not be involved at all in this computer leap from gene to structure, but the pendulum has reversed direction recently with the realization (particularly by the biotechnology industry) that protein chemistry is still centre stage in the understanding of physiological processes. The pathway by which structure–function relationships are revealed may make us uneasy at times, but it still involves proteins. We still need to know that Glu → Val is the sickle-cell haemoglobin mutation if we want to understand why insoluble material is formed in the cell.

Nevertheless, In a very real sense, the genetic code and its use as primary data has signalled the end of an era in protein science and the beginning of another. A future history, perhaps written by one of the bright new graduate students now entering this burgeoning field, will have to take over from where we left off.

Notes and references

Chapter 1

1. Söderbaum, H. G. (ed.) (1916). *Jac. Berzelius Lettres*. Uppsala. Vol. 5 is devoted to correspondence with G. J. Mulder.
2. English translation: 'The name protein which I propose to you for the organic oxide of fibrin and of albumin, I wanted to derive it from πρωτειοζ, because it seems to be the original or principal substance of animal nutrition.'
3. Smeaton, W. A. (1962). *Fourcroy, chemist and revolutionary, 1755–1809*. Heffer, Cambridge. Fourcroy worked closely with Antoine Lavoisier in establishing analytical methods for content of the elements that Lavoisier had defined. After John Dalton's *New system of chemical philosophy* was published, beginning in 1808, 'elements' would have been equated with 'atoms'.
4. Holmes, F. L. (1963). Elementary analysis and the origins of physiological chemistry. *Isis* 54, 50–81.
5. Holmes, F. L. (1964). Introduction to reprint of J. Liebig (1842). *Animal chemistry or organic chemistry in its application to physiology and pathology* (trans. W. Gregory), Cambridge. Reprinted in 1964 by Johnson Reprint Corp., New York, pp. 7–116.
6. Fruton, J. S. (1972). *Molecules and life*, Wiley-Interscience, New York, 1972, pp. 87–179. An updated version was published by Yale University Press in 1999, with the title, *Proteins, enzymes, genes*.
7. Coley, N. G. (1996). Studies in the history of animal chemistry and its relation to physiology. *Ambix* 43, 164–187.
8. Benfey, O. T. (1992). *From vital force to structural formulas*. Beckman Center for the History of Chemistry, Philadelphia. This is a brief summary, avoiding most technical difficulties. References to the original publications of Kekulé and others are provided.
9. Vickery, H. B. (1950). The origin of the word 'protein'. *Yale Journal of Biology and Medicine* 22, 387–393.
10. Jorpes, J. E. (1970). *Jac. Berzelius, his life and work* (trans. B. Steele). Almqvist & Wiksell, Stockholm.
11. Letters exchanged with Wöhler had been published earlier than the collection in ref. 1. See Wallach, O. (ed.) (1901). *Briefwechsel zwischen J. Berzelius und F. Wöhler*, 2 vols, Engelmann, Wiesbaden.
12. Even parcel post was still in the future. A letter from Mulder in 1836 tells Berzelius that he has been to the docks to arrange delivery of some books to him with the captain of a ship about to sail from Amsterdam to Stockholm—expected journey time was 20 days! (Letters were normally sent by courier.)
13. Mulder, G. J. (1837). Untersuchung mehrerer animalischer Stoffe, wie Fibrin, Eiweiss, Gallerte u. dgl. *Annalen der Pharmacie* 24, 256–265.
14. Mulder, G. J. (1838). Zusammensetzung von Fibrin, Albumin, Leimzucker, Leucin u.s.w. *Annalen der Pharmacie* 28, 73–82.
15. Brock, W. H. (1997). *Justus von Liebig: The chemical gatekeeper*. Cambridge University Press, Cambridge.

16. Liebig, J. (1839). *Instructions for the analysis of organic bodies*. Griffin & Tegg, Glasgow and London. Like most of Liebig's books, it was published simultaneously in German and in English translation. (References here and below are to the translations.)
17. Liebig, J. (1840). *Organic chemistry in its applications to agriculture and physiology* (trans. L. Playfair). Taylor & Walton, London.
18. Liebig, J. (1842). *Animal chemistry or organic chemistry in its application to physiology and pathology* (trans. W. Gregory), pp. 7–116. Cambridge. Reprinted in 1964 by the Johnson Reprint Corp., New York.
19. Glas, E. (1975). The protein theory of G. J. Mulder (1802–1880). *Janus* 62, 289–308.
20. The journal was renamed *Annalen der Chemie und Pharmacie* in 1840.
21. Anon (1839). Das enträthselte Geheimniss der geistigen Gährung. *Annalen der Pharmacie* 29, 100–104. For an English translation and commentary, see de Mayo, P., Stoesl, A., and Usselman, M. C. (1990). The Liebig-Wöhler satire on fermentation. *Journal of Chemical Education* 67, 552–553.
22. Snelders, H. A. M. (ed.) (1986). *The letters of Gerrit Jan Mulder to Justus von Liebig (1836–1846)*, published as *Janus Supplements* IX.
23. Snelders, ref. 22, p. 70.
24. Glas, E. (1976). The Liebig–Mulder controversy. On the methodology of physiological chemistry. *Janus* 63, 27–46.
25. Liebig, J. (1841). Ueber die stickstoffhaltigen Nahrungsmittel des Pflanzenreichs. *Annalen der Chemie und Pharmacie* 39, 129–160.
26. Ref. 18, p. 42
27. Laskowski, N. (1846). Ueber die Proteintheorie. *Annalen der Chemie und Pharmacie* 58, 129–166.
28. Fleitmann, T. (1847). Ueber die Existenz eines schwefelfreien Proteins. *Annalen der Chemie und Pharmacie* 61, 121–126. Fleitmann in his text called it the 'Muldersche Protein', and went to great length to prove that sulphur was a genuine 'Verbindung' and not a contaminant.
29. We know today, of course, that sulphur is an intrinsic component of two constituent amino acids and, as such, should not be removable.
30. Dumas, J. B. and Cahours, A. (1842). Sur les matières azotées neutres de l'organisation: *Annales de Chimie et Physique* 6, 385–448.
31. Liebig, J. (1847). Ueber die Bestandteile der Flüssigkeiten des Fleisches. *Annalen der Chemie und Pharmacie* 62, 257–369. The criticism of Mulder is on pp. 268–270.
32. Mann, G. (1906). *Chemistry of the proteids*. Macmillan, London.
33. Dyke, G. V. (1993). *John Lawes of Rothamsted*. Hoos Press, Harpenden.
34. The existence of different atomic weight scales did not create serious problems before concepts of valency and interatomic bonding became established. The famous Karlsruhe conference, which was forced to adopt different sets of atomic weights for inorganic and organic chemistry, did not take place until 1860. See Partington, J. R. (1961–1964). *History of chemistry*, vol. 4, pp. 166, 488. Macmillan, London.

Chapter 2

1. Osborne, T. B. (1892). Crystallized vegetable proteids. *American Chemical Journal* 14, 662–689.
2. Haüy, R.-J. (1801). *Traité de minéralogie*. Louis Libraire, Paris. A second edition was published in 1822 and *Traité de crystallographie* in the same year.
3. Kühne, W. (1859). Untersuchungen über Bewegungen und Veränderungen der con-

tractilen Substanzen. *Archiv für Anatomie und Physiologie* 748–835; Kühne, W. (1864). *Untersuchungen über das Protoplasma und die Contractilität.* W. Engelmann, Leipzig.

4. See Chapter 5, note 18.

5. Hünefeld, F. L. (1840). *Der Chemismus in der thierischen Organisation,* pp. 158–163. Brockhaus, Leipzig.

6. Preyer, W. T. (1871). *Die Blutkrystalle.* Mauke's Verlag, Jena. Preyer was born in Manchester, but moved to Germany when he was sixteen. His most significant work is said to have been in developmental psychology: a book, *Die Seele des Kindes* (Leipzig, 1882), was particularly influential and ran to eight editions.

7. Reichert, E. T. and Brown, A. P. (1909). *The differentiation and specificity of corresponding proteins and other vital substances in relation to biological classification and organic evolution. The crystallography of hemoglobin.* Washington, Carnegie Institution.

8. Kossel, A. (1901). Ueber den gegenwärtigen Stand der Eiweisschemie. *Berichte der deutschen chemischen Gesellschaft* 34, 3214–3245. Kossel is a major figure in the history of proteins. He was someone who always expressed his opinions clearly and forcibly. There was no question in his mind about the macromolecular nature of proteins, demonstrated unambiguously in his opinion by serial degradation to successively smaller products. Similarly, he argued that ability to absorb other molecules in the crystalline state made it 'not permissible' to regard a protein as pure, simply on the basis of its having been crystallized.

9. Vaubel, W. (1899). Ueber die Molekulargrösse der Eiweisskörper. *Journal für praktische Chemie* 60, 55–71.

10. Osborne, T. B. (1892). Crystallized vegetable proteids. *American Chemical Journal* 14, 662–689.

11. Pirie, N. W. (1979). Purification and crystallization of proteins. *Annals of the New York Academy of Sciences* 325, 21–32. Pirie was an agricultural biochemist.

12. Edsall, J. T. (1972). Blood and hemoglobin: the evolution of knowledge of functional adaptation in a biochemical system. *Journal of the History of Biology* 5, 205–257.

13. Teichmann, L. (1853). Ueber die Krystallisation der organischen Bestandtheile des Bluts. *Zeitschrift für rationelle Medicin* NF3, 375–388. Introduces the word 'haemin'.

14. Hoppe-Seyler, F. (1864). Ueber die optischen und chemischen Eigenschaften des Blutfarbstoffs. *Virchows Archiv für Pathologie, Anatomie und Physiologie* 29, 233–235, 597–600.

15. Hoppe, F. (1862). Ueber das Verhalten des Blutfarbstoffs im Spectrum des Sonnenlichtes. *Archiv für pathologische und anatomische Physiologie* 23, 446–499. Hoppe did not change his name to Hoppe-Seyler until after this paper was published. Cf. Edsall, ref. 12, p. 209.

16. Stokes, G. G. (1864). On the reduction and oxidation of the colouring matter of blood. *Proceedings of the Royal Society* 13, 355–364.

17. Pauling, L. and Coryell, C. D. (1936). The magnetic properties and structure of hemoglobin, oxyhemoglobin and carbonmonoxyhemoglobin. *Proceedings of the National Academy of Sciences of the USA* 22, 210–216.

18. Gamgee, A. (1898). Haemoglobin: its compounds and the principal products of its decomposition. In Schäfer, E. A. (ed.), *Textbook of physiology,* vol. 1, pp. 185–260. Y. J. Pentland, Edinburgh and London.

19. Barcroft, J. (1914) *The respiratory function of the blood.* Cambridge University Press, Cambridge.

20. See ref. 1 for a review. Specific preparative procedures are given in separate papers: Osborne, T. B. (1892). *American Chemical Journal* 14, 212–224, 629–661.

21. Hofmeister, F. (1890–1892). Über die Darstellung von kristallisierten Eieralbumin und die Kristallisierbarkeit kolloider Stoffe. *Zeitschrift für physiologische Chemie* **14**, 165–172; Ueber die Zummansetzung des kristallisierten Eieralbumins. Ibid **16**, 187–191.

22. Hopkins, F. G. and Pinkus, S. N. (1898). Observations on the crystallization of animal proteids. *Journal of Physiology* **23**, 130–136.

23. Hopkins, F. G. (1900). On the separation of a pure albumin from egg-white. *Journal of Physiology* **25**, 306–330.

Chapter 3

1. Hofmeister, F. (1902). Ueber Bau und Gruppierung der Eiweisskörper. *Ergebnisse der Physiologie* **1**, 759–802. The quoted extract appears on p. 792 of this long, detailed paper. The translation is from Teich, M. with Needham, D. M. (1992). *A documentary history of biochemistry 1770–1940.* Leicester University Press, Leicester. The *italics* are ours.

2. Fischer, E. (1906). Untersuchungen über Aminosäuren, Polypeptide und Proteine. *Berichte der deutschen chemichen Gesellschaft* **39**, 530–610. See also Fischer, E. (1922). In M. Bergmann (ed.) *Gesammelte Werke.* Springer, Berlin. Reprinted in 1987.

3. Cohn, E. J. (1925).The physical chemistry of the proteins. *Physiological Reviews* **5**, 349–437.

4. Vickery, H. B. and Schmidt, C. L. A. (1931). The history of the discovery of the amino acids. *Chemical Reviews* **9**, 169–318.

5. Vickery, H. B. (1972).The history of the discovery of the amino acids. II. A review of amino acids described since 1931 as components of native proteins. *Advances in Protein Chemistry* **26**, 81–171. This review was written *after* the biosynthetic mechanism and the code that links amino acids to DNA triplets were discovered. It includes iodine-containing amino acids and others that we would now call post-translational modifications.

6. Hlasiwetz, H. and Habermann, J. (1871–1873). Ueber die Proteinstoffe. *Annalen der Chemie* **159**, 304–333; **169**, 150–166. These papers by two relatively obscure Czechs have influenced several aspects of protein chemistry and are considered by many to be 'classics'. See Chapter 4, p. 45.

7. Asparagine had actually been identified as a chemical substance and given its name in 1806, long before its protein origin could even be guessed at. It had been obtained from extracts of asparagus shoots and its early appearance in the chemical roster is due to its ready crystallization (as a monohydrate) even from complex mixtures and, hence, its ready availability as a pure substance.

8. Benfey, O. T. (1992). *From vital force to structural formulas.* Beckman Center for the History of Chemistry, Philadelphia.

9. Koerting, W. (1968). Die Deutsche Universität in Prag. Die letzten hundert Jahre ihrer medizinischen Fakultät. Munich.

10. Bochalli, R. (1948). Die Gesellschaft deutscher Naturforscher und Ärzte als Spiegelbild der Naturwissenschafter und der Medizin. *Naturwissenschaftliche Rundschau (Stuttgart)* **1**, 275–278. Querner, H. and Schipperges, H. (eds) (1972). *Wege der Naturforschung, 1822–1972,* Springer, Berlin. Querner, H. (ed.) (1972). *Schriftenreihe zu Geschichte der Versammlung deutscher Naturforscher und Ärzte.* Gerstenberg, Hildesheim.

11. The first meeting was in 1822 in Leipzig. Meetings were held annually thereafter, with a handful of exceptions, for example, 1848, which was a year of national revolution in

Germany. Berzelius was a plenary speaker in 1828, Liebig in 1825, Helmholtz in 1857, etc. Neither Mulder nor Kekulé is on the list of plenary speakers.

12. Fruton, J. S. (1985). Contrasts in scientific style. Emil Fischer and Franz Hofmeister: their research groups and their theory of protein structure. *Proceedings of the American Philosophical Society* **129**, 313–370; Fruton, J. S. (1990). *Contrasts in scientific style: research groups in the chemical and biochemical sciences.* American Philosophical Society, Philadelphia.

13. Vickery, H. B. and Osborne, T. B. (1928). A review of hypotheses of the structure of proteins. *Physiological Reviews* **8**, 393–446.

14. Hoesch, K. (1921). *Emil Fischer, sein Leben und sein Werk.* Verlag Chemie, Berlin.

15. Forster, M. O. (1920). Emil Fischer memorial lecture. *Journal of the Chemical Society* **117**, 1157–1201.

16. Ostwald, W. (1927). *Lebeslinien.* Klasing, Berlin. Cited by Fruton, ref. 12.

17. Twenty years later, Fischer's mentor, Adolf von Baeyer, was claiming that organic chemistry itself was exhausted. It is a wonder that chemists have found anything to do for the last century.

18. Moves of this kind were typical of the jockeying for positions in German academia at the time, another consequence of prosperity and expanding opportunities.

19. Fischer, E. (1894). Einfluss der Konfiguration auf die Wirkung der Enzyme. I. *Berichte der deutschen chemischen Gesellschaft* **27**, 2985–2993.

20. The letter is quoted by Fruton, loc. cit. ref. 12.

21. Fischer, E. and Fourneau, E. (1901). Über einige Derivate des Glykocolls. *Berichte der deutschen chemischen Gesellschaft* **34**, 2868–2877.

22. Forster's memorial lecture, ref. 15, loc. cit., p. 1184.

23. Fischer, E. (1902). Über die Hydrolyse der Proteinstoffe. *Chemiker Zeitung* **26**, 939–940.

24. Pohl, J. and Spiro, K. (1923), Franz Hofmeister, sein Leben und Wirken. *Ergebnisse der Physiologie* **22**, 1–50. There is also a well-written obituary by Neuberg, C. (1923). *Biochemische Zeitschrift* **134**, 1–2, which is brief but informative.

25. It happens that volume 127 of *Biochemische Zeitschrift* (1922) was a 'Festband' for Hofmeister, published just a few weeks before he died.

26. Hofmeister, F. (1889). Über die Darstellung von kristallisierten Eieralbumin und die Kristallisierbarkeit kolloider Stoffe. *Zeitschrift für physiologische Chemie* **14**, 165–172.

27. Hofmeister, F. (1901). *Die chemische Organisation der Zelle.* Vieweg, Braunschweig.

28. Hofmeister, F. (1902). Ueber den Bau des Eiweissmoleküls. *Naturwissenschaftliche Rundschau (Braunschweig)* **17**, 529–533, 545–549. This is the abstract version.

29. Hofmeister, F. (1902). Ueber Bau und Gruppierung der Eiweisskörper. *Ergebnisse der Physiologie* **1**, 759–802.

30. Fischer, E. (1906). See ref. 2 above.

31. Fischer, E. (1907). Synthetical chemistry in its relation to biology (Faraday Lecture). *Journal of the Chemical Society* **91**, 1749–1765. Fischer was originally invited to give a Faraday lecture in 1895, but could not accept because of poor health. A lecture in 1895 would have been relatively uninteresting because that was before he had begun his work with amino acids or peptides.

32. Hofmeister, F. (1908). Einiges über die Bedeutung und den Abbau der Eiweisskörper. *Archiv für experimentelle Pathologie und Pharmacologie* (1908 Supplement), 273–281.

33. Kossel, A. (1901). Ueber den gegenwärtigen Stand der Eiweisschemie. *Berichte der deutschen chemischen Gesellschaft* **34**, 3214–3245.

34. Kossel, A. (1912). Lectures on the Herter Foundation. *Johns Hopkins Hospital Bulletin* 23, 65–76. This paper speculates about a concept close to specific amino acid sequences.

35. Vickery, H. B. and Osborne, T. B. (1928). Loc. cit. ref. 13.

36. See, for example, Plimmer, R. H. A. (1908). *The chemical constitution of the proteins.* Longmans, London, which is dedicated to Emil Fischer, 'the master of organic chemistry in its relation to biology'.

37. Abderhalden, E. (1924). Diketopiperazines. *Naturwissenschaften* 12, 716.

38. Abderhalden, E. and Komm, E. (1924). Über die Anhydridstruktur der Proteine. *Zeitschrift für physiologische Chemie* 139, 181–204.

39. Wrinch, D. M. (1937) On the pattern of proteins. *Proceedings of the Royal Society* A160, 59–86; Wrinch D. M. (1937). The cyclol hypothesis and the 'globular' proteins. *Proceedings of the Royal Society* A161, 505–524.

Chapter 4

1. This chapter is based on an earlier article: Tanford, C. and Reynolds, J. (1999). Protein chemists bypass the colloid/macromolecule debate. *Ambix* 46, 33–51.

2. Zinoffsky, O. (1885). Ueber die Grösse des Hämoglobinmolecüls. *Zeitschrift für physiologische Chemie* 10, 16–34.

3. Flory, P. J. (1953). *Principles of polymer chemistry*, pp. 3–28. Cornell University Press, Ithaca.

4. Morawetz, H. (1985). *Polymers: The origins and growth of a science.* Wiley, New York.

5. Morris, P. J. T. (1986). *Polymer pioneers.* Center for History of Chemistry, Philadelphia.

6. Olby, R. (1970). The macromolecular concept. *Journal of Chemical Education* 47, 168–174.

7. Furukawa, Y. (1998). *Inventing polymer science.* University of Pennsylvania Press, Philadelphia.

8. Staudinger, H. (1961). *Arbeitserinerungen.* Hüthig, Heidelberg. English translation entitled *From organic chemistry to macromolecules.* Wiley, New York, 1970.

9. Florkin, M. and Stotz, E. H. (eds) (1972). *Comprehensive biochemistry*, vol. 32, pp. 279–284. Elsevier, Amsterdam. The Florkin–Stotz position is substantially modified—protein chemists given much more credit—in a subsequent volume of the same treatise: Laszlo, P. (1986). The notion of macromolecules, in vol. 34A, pp. 12–22.

10. Words spoken in frustration at a lecture in 1925 at the Zürich Chemical Society. See Morawetz, op. cit. ref. 4, p. 86.

11. See explicitly the table reproduced in Fig. 1.2 in Chapter 1.

12. Thudichum, J. L. W. (1872). *A manual of chemical physiology.* Longmans, London.

13. Hüfner, G. (1894). Neue Versuche zur Bestimmung der Sauerstoffcapacität des Blutfarbstoffs. *Archiv für Physiologie*, 130–176.

14. Kossel, A. and Kutscher, F. (1900). Beiträge zur Kenntniss der Eiweisskörper. *Zeitschrift für physiologische Chemie* 31, 165–214.

15. Hofmeister, F. (1902). Ueber Bau und Gruppierung der Eiweisskörper. *Ergebnisse der Physiologie* 1, 759–802.

16. Hlasiwetz, H. and Habermann, J. (1871–1873). Ueber die Proteinstoffe. *Annalen der Chemie* 159, 304–333; 169, 150–166.

17. Kossel, A. (1901). Ueber den gegenwärtigen Stand der Eiweisschemie. *Berichte der deutschen chemischen Gesellschaft* 34, 3214–3245.

18. Barth, L. (1876). Nekrologie auf Heinrich Haslewitz. *Berichte der deutschen chemischen Gesellschaft* 9, 1961-1992. See also McConnell, V. F. (1953). Hlasiwetz and Barth— pioneers in the structural aspects of plant products. *Journal of Chemical Education* 30, 380-385. Hlasiwetz's colleague, Josef Haberman (1841-1914), spent most of his career at the University of Brno.

19. Servos, J. W. (1990). *Physical chemistry from Ostwald to Pauling.* Princeton University Press; Laidler, K. J. (1993). *The world of physical chemistry.* Oxford University Press, New York.

20. Sabanjeff, A. (1891). Kryoskopische Untersuchung der Kolloide. Bestimmung des Molekulargewichtes von Kolloiden nach Raoult's Methode. *Chemisches Centralblatt* 62, part 1, 10-12. (Abstract of an earlier Russian paper.)

21. Kauffman, G. B. (1972). Biography of Thomas Graham. In *Dictionary of scientific biography*, vol. 5.

22. Graham, T. (1861). Liquid diffusion applied to analysis. *Philosophical Transactions of the Royal Society* 151, 183-224. Graham quit his academic appoinment at University College (London) in 1854 to become Master of the Mint, the last to hold that title, Isaac Newton having been the first. His 1861 paper dates from this late period in his career.

23. Kekulé, A. (1878). Inaugural lecture given at the University of Bonn. *Nature* 18, 210.

24. Ostwald, W. (1884-1885). *Lehrbuch der allgemeinen Chemie*, vol. 1, p. 527. W. Engelmann, Leipzig.

25. van Bemmelen, J. M. (1888). Sur la nature des colloides et leur teneur en eau. *Recueil des Travaux Chimiques des Pays-Bas* 7, 37-68. For English translation see Hatschek, E. (ed.) (1925). *The foundations of colloid chemistry.* Ernest Benn, London.

26. Picton, H. and Linder, S. E. (1892). Solution and pseudo-solution. Part I. *Journal of the Chemical Society* 61, 148-172. See also Part III, which was published in 1897 in *Journal of the Chemical Society* 71, 568-573.

27. Neither Picton nor Linder kept on with their research after these papers were published, but they continued to be held in high esteem because their work laid the foundation for electrophoresis as an experimental tool. After World War I Picton became active in the peace movement and vigorously promoted good relations with Germany. In 1923 he was invited to address the 2nd general meeting of the Kolloidgesellschaft in Jena, where he gave a moving talk, describing himself as a 'phantom of the past'. See Report of the second general meeting of the Kolloidgesellschaft (1923). *Kolloid Zeitschrift* 33, 257-261. Also Hatschek, E. (ed.) (1926). Klassische Arbeiten über kolloide Lösungen, *Ostwald's Klassiker der Exakten Wissenschaften*, vol. 217.

28. Servos, J. W. (1990). op. cit. ref. 19, ch. 3; Hager, T. (1995). *Force of nature: the life of Linus Pauling*, pp. 77-84. Simon & Schuster, New York.

29. Noyes, A. A. (1905). The preparation and properties of colloidal mixtures. *Journal of the American Chemical Society* 27, 85-104.

30. Lumière, A. (1921). *Role des colloides chez les êtres vivants, essai de biocolloidologie.* Masson et Cie, Paris.

31. Letter from Jacques Loeb to L. Michaelis, Jan 27, 1921, in Werner. P. (1996). *Otto Warburg's Beitrag zur Atmungstheorie.* Basiliken-Presse, Marburg.

32. Servos, J. W. (1990). *Physical chemistry from Ostwald to Pauling*, ref. 19, pp. 299-324.

33. Laidler, K. J. (1993). *The world of physical chemistry*, ref. 19, pp. 48-52, 292-293.

34. Pauly, P. (1987). *Controlling life: Jacques Loeb and the engineering ideal in biology.* Oxford University Press, Oxford.

35. First published in 1906 as *Zeitschrift für Chemie und Industrie der Kolloide*. Ostwald became editor in 1907.

36. Ostwald, Wo. (1907). Zur Systematik der Kolloide. *Zeitschrift für Chemie und Industrie der Kolloide (Kolloid Z.)* 1, 291–300, 331–341. It is customary to refer to Wolfgang Ostwald by use of 'Wo' as an initial, reserving the initial 'W' for his father Wilhelm.

37. Ostwald, G. (1953). *Wilhelm Ostwald, mein Vater*. Berliner Union, Stuttgart.

38. Needham, J. and Baldwin, E. (eds) (1949). *Hopkins and biochemistry*, p. 179. Heffer, Cambridge.

39. Hausmann, W. (1900). Ueber die Vertheilung des Stickstoffs von Eiweissmolekul. *Zeitschrift für physiologische Chemie* 29, 136–145.

40. Jones, M. E. (1953). Albrecht Kossel, a biographical sketch. *Yale Journal of Biology and Medicine* 26, 80–97. Kossel won a Nobel Prize in 1910 for his work on proteins and 'nucleic substances'.

41. Kossel, A. and Kutscher, F. (1900). Beiträge zur Kenntniss der Eiweisskörper. *Zeitschrift für physiologische Chemie* 31, 165–214.

42. Kossel, A. (1912). Lectures on the Herter Foundation. *Johns Hopkins Hospital Bulletin* 23, 65–76. The first actual determination of an amino acid sequence had to wait until Sanger's work on insulin in 1953.

43. Vickery, H. B. (1931). Biographical memoir of T. B. Osborne. *Biographical Memoirs of the National Academy of Sciences* 14, 261–303.

44. Osborne, T. B. (1902). Sulphur in protein bodies. *Journal of the American Chemical Society* 24, 140–167.

45. Osborne, T. B. and Harris, I. F. (1903). Nitrogen in protein bodies. *Journal of the American Chemical Society* 25, 323–353.

46. Osborne, T. B. (1909). *The vegetable proteins*. Longmans, London.

47. Holter, H. and Møller, M. (eds) (1976). *The Carlsberg Laboratories 1876/1976*. Rhodos, Copenhagen.

48. Sørensen, S. P. L. (1909). Enzyme studies II. Measurement and significance of hydrogen ion concentration in enzyme processes. *Comptes rendus des traveaux du Laboratoire Carlsberg* 8, 1–168 (in French); *Biochemische Zeitschrift* 21, 131–304 (in German).

49. Güntelberg, A. V. and Linderstrøm-Lang, K. (1949). Osmotic pressure of plakalbumin and ovalbumin solutions. *Comptes rendus des traveaux du Laboratoire Carlsberg, Sér. chim.* 27, 1–25.

50. Sørensen, S. P. L., Høyrup, M. and others (1915–1917). Studies on proteins. *Comptes rendus des traveaux du Laboratoire Carlsberg* 12, 1–372. Important papers in this series include Sørensen, S. P. L. and Høyrup, M. On the state of equilibrium beteen crystallised egg albumin and surrounding mother liquor, and on the application of Gibbs' phase rule to such systems, ibid. 12, 213–261, and Sørensen, S. P. L. On the osmotic pressure of egg albumin solutions, ibid. 12, 262–371.

51. Barcroft, J. and Hill, A. V. (1910). The nature of oxyhaemoglobin, with a note on its molecular weight. *Journal of Physiology* 39, 411–428.

52. William Hardy (1864–1934) made many valuable contributions to the study of the electrical properties of proteins. On the subject of protoplasm see Hardy, W. B. (1899). Structure of cell proptoplasm. *Journal of Physiology* 24, 158–210.

53. Abderhalden, E. (1922). Ueber die Beziehung der Kolloidchemie zur Physiologie. *Kolloid Zeitschrift* 31, 276–279. The paper was given at the first meeting of the Kolloid-Gesellschaft.

54. Abderhalden, E. and Komm, E. (1924). Über die Anhydridstruktur der Proteine. *Zeitschrift für physiologische Chemie* 139, 181–204.

55. For other Abderhalden aberrations see Deichmann, U. and Müller-Hill, B. (1998). The fraud of Abderhalden's enzymes. *Nature* 393, 109–111.

56. Fischer, E. (1913). Synthese von Depsiden, Flechtenstoffen und Gerbstoffen. *Berichte der deutschen chemischen Gesellschaft* 46, 3253–3289. The paper is the text of a lecture delivered in Vienna in September 1913.

57. Another puzzling backslider late in life was S. P. L. Sørensen, who (in 1930!) appears to disown his earlier (1915) criticism of colloid chemistry. Edsall has discussed the experimental fallacy that led to this, but no one has explained why Sørensen chose to ignore all his own previous definitive experiments on the subject. See Sørensen, S. P. L. (1930). The constitution of soluble proteins as reversibly dissociable component systems. *Comptes rendus des travaux du Laboratoire Carlsberg* 18, no. 5, 1–124; Cohn, E. J. and Edsall, J. T. (1943). *Proteins, amino acids and peptides as ions and dipolar ions*, pp. 576–585. Reinhold, New York.

58. Cohnheim, O. (1900). *Chemie der Eiweisskoerper*. Vieweg, Braunschweig. (A second edition was published in 1904); Mann, G. (1906). *Chemistry of the proteids*. Macmillan, London.

59. Plimmer, R. H. A. (1908). *The chemical constitution of the proteins*. Longmans, London. Schryver, S.B. (1909). *The general characters of the proteins*. Longmans, London. Both authors were on the faculty at University College, London.

60. Robertson, T. B. (1912). *Die physikalische Chemie der Proteine*. Steinkopff, Dresden. Robertson, T. B. (1918). *The physical chemistry of the proteins*. Longmans, New York. The book was published as a German translation six years prior to its first publication in the United States.

61. Abderhalden, E. (1906). *Lehrbuch der physiologischen Chemie in dreissig Vorlesungen*. Urban & Schwarzenberger, Berlin; Abderhalden, E. (1908). *Text-book of physiological chemistry in thirty lectures*. Wiley, New York.

62. Pauli, W. (1902). *Der kolloidale Zustand und die Vorgänge in der lebendigen Substanz*, Vieweg, Braunschweig; Pauli, W. (1922). *Colloid chemistry of proteins*. J. & A. Churchill, London.

63. Svedberg, T. (1909). *Herstellung kolloider Lösungen*. Steinkopff, Dresden; Svedberg, T. (1921). *The formation of colloids*. J. & A. Churchill, London.

64. Einstein, A. (1907), Theoretical observations on the Brownian motion. *Zeitschrift für Elektrochemie* 13, 41–42. This paper is reprinted in Einstein's popular book on this subject: Einstein, A. (1926). *Investigations on the theory of the Brownian movement*. Methuen, London. (Paperback edition published by Dover, New York, in 1956.)

65. Readers with short memories may want to be reminded that so-called 'Brownian motion' was named after the Scottish botanist Robert Brown, who first reported in 1828 that pollen grains suspended in a liquid spontaneously undergo tiny irregular movements. Einstein's quantitative statistical explanation of this phenomenon as arising from random bombardment of the pollen grains by independent particles of liquid was decisive in convincing many people that independent 'molecules' really exist. Einstein related Brownian motion directly to osmotic pressure and to diffusion: the subject became in his hands a theoretical core for the molecular-kinetic theory of heat and matter and its subsequent applications to a host of tools—ultracentrifugation, viscosity, etc.—for characterization of protein molecules. What is important here in the context of this book is to appreciate that protein science was at the heart of fundamental aspects of theoretical physics and that protein *chemists* (through Svedberg's work and its consequences) would in this respect have had a more sophisticated view of the

molecular world than most other organic chemists. Note the contrast to optical asymmetry, mentioned earlier (footnote on p. 31) as a subject where general organic chemists were notably proficient.

66. Svedberg published a detailed account of his conversion to the macromolecular concept for the colloid community in the colloid scientist's own journal, *Kolloid Zeitschrift*. Svedberg, T. (1930). Ultrazentrifugale Dispersitätsbestimmungen an Eiweisslösungen. *Kolloid Zeitschrift* 51, 10–24.

67. Svedberg, T. and Fåhraeus, R. (1926). A new method for the determination of the molecular weight of proteins. *Journal of the American Chemical Society* 48, 430–438.

68. Svedberg, T. (1929). Mass and size of protein molecules. *Nature* 123, 871; Svedberg, T. and Pedersen, K.O. (1940). *The ultracentrifuge*. Clarendon, Oxford.

69. Söderbaum, H.G. (1966). Presentation speech for award to T. Svedberg. In *Nobel lectures—Chemistry*. Elsevier, Amsterdam. The date of the award and presentation speech was 1926.

70. See programme in *Mitteilungen der Gesellschaft deutscher Naturforscher und Ärzte* 3, no. 12 (1926).

71. Papers presented at the meeting were published: Waldschmidt-Leitz, E. (1926). Zur Struktur der Proteine. *Berichte der deutschen chemischen Gesellschaft* 59, 3000–3007; Bergmann, M. (1926). Allgemeine Stukturchemie der komplexen Kohlenhydrate and der Proteine. *Berichte der deutschen chemischen Gesellschaft* 59, 2973–2981. The work of his own that Bergmann chiefly cites is not about proteins, but about inulin, a polysaccharide.

72. Waldschmidt-Leitz, E. (1933). The chemical nature of enzymes. *Science* 78, 189–190.

73. Staudinger, H. (1926). Die Chemie der hochmolekularen organischen Stoffe im Sinne der Kekuléschen Strukturlehre. *Berichte der deutschen chemischen Gesellschaft* 59, 3019–3043.

74. Svedberg, T. (1966). In *Nobel lectures—chemistry*. Elsevier, Amsterdam. The award was made in 1926. The physics prize in the same year also went to a colloid-oriented scientist, namely Jean Perrin.

75. Staudinger, H. (1966). In *Nobel lectures—chemistry*. Elsevier, Amsterdam. Staudinger's Nobel Prize for chemistry came rather tardily in 1953.

Chapter 5

1. Cohn, E. J. and Prentiss, A. M. (1927). Studies in the physical chemistry of the proteins. VI. The activity coefficients of the ions in certain oxyhemoglobin solutions. *Journal of General Physiology* 8, 619–639.

2. See, for example, Laidler, K. J. (1995). *The world of physical chemistry* (revised edition), pp. 207–227. Oxford University Press, New York.

3. Ostwald, W. (1887). *Lehrbuch der Allgemeinen Chemie*, vol. 2, pp. 821–829. W. Engelmann, Leipzig.

4. Loeb, J. (1898). The biological problems of today: physiology. *Science* 7, 154–156.

5. Loeb, J. (1922). *Proteins and the theory of colloidal behavior*. McGraw-Hill, New York. (2 edn, 1924).

6. Bredig, G. (1894). Über die Affinitätsgrossen der Basen. *Zeitschrift für physikalische Chemie* 13, 289–326; especially specific footnote on p. 323. Bredig subsequently became a colleague of Walther Nernst and famous in his own right for thermodynamic

studies at low temperatures and other work. For a biography, see W. Kuhn (1962). Nachruf auf G. Bredig. *Chemische Berichte* **95**, xlvii–lvii.

7. Winkelblech, K. (1901). Über amphotere Elektrolyte und innere Salze. *Zeitschrift für physikalische Chemie* **36**, 546–595.

8. Küster, F. W. (1897). Kritische Studien zur volumetrischen Bestimmung von karbon-athaltigen Alkalilaugen und von Alkalicarbonaten, sowie über das Verhalten von Phenolphtalein und Methylorange als Indikatoren. *Zeitschrift für inorganische Chemie* **13**, 127–150.

9. Edsall, J. T. (1936/37). Raman spectra of amino acids and related compounds. *Journal of Chemical Physics* **4**, 1–8; **5**, 225–237, 508–517.

10. A good inside view of this confusion can be obtained by reading the proceedings of a meeting of the German Electrochemical Society held in Göttingen in 1899, where a brief talk by Bredig was heatedly discussed by Ostwald, Küster, and others: see Bredig, G. (1899). Über amphotere Elektrolyte und innere Salze. *Zeitschrift für Elektrochemie* **6**, 33–37.

11. Adams, E. Q. (1916). Relations between the constants of dibasic acids and of ampho-teric electrolytes. *Journal of the American Chemical Society* **38**, 1503–1510. Adams (1888–1971) received his Ph.D. in 1914 under the direction of the great physical chemist G. N. Lewis at the University of California. This was the period when G. N. Lewis was interested in chemical bonding and spectroscopy, and Adams became in-volved with acids and bases because of the huge effects of molecular charge on spectra. For a brief biography see Tarbell, D. S. (1990). *Journal of Chemical Education* **67**, 7–8.

12. Bjerrum, N. (1923). Die Konstitution der Ampholyte, besonders der Aminosäuren, und ihre Dissociationkonstanten. *Zeitschrift für physikalische Chemie* **104**, 147–173.

13. Picton, H. and Linder, S. E. (1892). Solution and pseudo-solution. Part I. *Journal of the Chemical Society* **61**, 148–172. See also Part III, which was published in 1897 in *Journal of the Chemical Society* **71**, 568–573.

14. Given the pivotal place of this work in our history, its importance never diminished by later perspectives, it is rather surprising that neither Picton nor Linder kept on with their research after these papers were published. See Chapter 4, ref. 27.

15. Hardy, W. B. (1899). On the coagulation of proteid by electricity. *Journal of Physiology* **24**, 288–304.

16. Pauli, W. (1906). Untersuchungen über physikalische Zustandsänderungen der Kolloide. Die elektrische Ladung von Eiweiss. *Beiträge zur chemischen Physiologie und Pathologie* **7**, 531–547. The work was continued under a similar title by Pauli, W. and Handovsky, H. (1908). ibid. **11**, 415–448. Pauli was the father of the famous physicist who was responsible for the Pauli exclusion principle.

17. Michaelis, L. (1909). Die elektrische Ladung des Serumalbumins und der Fermente. *Biochemische Zeitschrift* **19**, 181–185; Michaelis, L. and Davidsohn, H. (1911). Der iso-elektrische Punkt des genuinen und des denaturierten Serumalbumins. *Biochemische Zeitschrift* **33**, 456–473.

18. All animal chemistry used to be disparaged by the dictum,[19] 'Tierchemie ist Schmierchemie'. Measurements at the level of accuracy of the pH values here cited grad-ually began to dispel that conception when it came to proteins. Furthermore, repro-ducibility of relatively small differences between isoelectric points of different proteins contributed greatly to respect for purity of individual protein preparations.

19. Hopkins, F. G. (1913). Address to the British Association. In Needham, J. and Baldwin, E. (eds) (1949). *Hopkins and biochemistry*, p. 137. Heffer, Cambridge.

20. Bugarszky, S. and Liebermann, L. (1898). Ueber das Bindungsvermögen eiweissartiger Körper für Salzsäure, Natriumhydroxyd und Kochsalz. *Pflüger's Archiv* 72, 51–74.

21. Nernst, W. (1889). Die elektromotorische Wirksamkeit der Ionen. *Zeitschrift für physikalische Chemie* 4, 129–181; (1894). Zur Dissociation des Wassers. *Zeitschrift für physikalische Chemie* 14, 155–156.

22. Osborne, T. B. (1902). The basic character of the protein molecule and the reaction of edestin with definite quantities of acids and alkalies. *Journal of the American Chemical Society* 24, 39–78.

23. Sørensen, S. P. L., Høyrup, M., and others (1915–1917). Studies on proteins. *Comptes rendus des travaux du Laboratoire Carlsberg* 12, 1–372.

24. Linderstrøm-Lang, K. (1924). On the ionisation of proteins. *Comptes rendus des travaux du Laboratoire Carlsberg* 15, no. 7, 1–29.

25. Cohn, E. J. (1925). The physical chemistry of proteins. *Annual Review of Physiology* 5, 350–437.

26. Loeb, J. (1922). *Proteins and the theory of colloidal behavior.* McGraw-Hill, New York, 1922; 2nd edn, 1924.

27. Michaelis, L. (1922). *Die Wasserstoffionenkonzentration*, 2nd edn. Springer, Berlin.

28. Michaelis, L. (1926). *Hydrogen ion concentration*, Vol. 1. *Principles of the theory.* Williams & Wilkins, Baltimore.

29. Abderhalden, E. and Komm, E. (1924). Über die Anhydridstruktur der Proteine. *Zeitschrift für physiologische Chemie* 139, 181–204. See also Chapter 3, p. 40.

30. Weber, H. H. (1930). Die Bjerrumsche Zwitterionentheorie und die Hydration der Eiweisskörper. *Biochemische Zeitchrift* 218, 1–35.

31. Holter, H. (1976). Biography of Kai Linderstrøm-Lang, in Holter, H. and Møller, K. M. (eds) (1976). *The Carslberg Laboratory* pp. 88–117, Rhodos, Copenhagen. There is an obituary, written by J. T. Edsall in 1959 in *Advances in Protein Chemistry* 14, xiii–xxiii.

32. The accepted molecular weight for egg albumin at the time was 34 000. Use of the later corrected value of 45 000 (ref. 49, Chapter 4) would have increased the radius to 24 Å.

33. Sutherland, W. (1905). A dynamical theory of diffusion for non-electrolytes and the molecular mass of albumin. *Philosophical Magazine* Ser. 6, 9, 781–785. An earlier example of Sutherland's genius was his proposal that strong electrolytes were completely dissociated into ions at all concentrations, contrary to the prevailing view—qualitatively introducing the concepts of the Debye–Hückel theory twenty years before Debye. (See Laidler, ref. 2, p. 216.) It may be noted that the title of the paper on particle shape suggests that Sutherland was not aware that proteins, too, must be electrolytes.

34. Sackur, O. (1902). Das elektrische Leitvermögen und die innere Reibung von Lösungen des Caseins. *Zeitschrift für physikalische Chemie* 41, 672–680.

35. Chick, H. and Lubrzynska, E. (1914). The viscosity of some protein solutions. *Biochemical Journal* 8, 59–69; Chick, H. The viscosity of protein solutions. II. Pseudo-globulin and euglobulin (horse). *Biochemical Journal* 8, 261–280. (Also see other papers by Chick and co-workers in *Biochemical Journal* 7 (1913) and by E. Hatschek in various journals between 1910 and 1913.)

36. Einstein, A. (1906). Eine neue Bestimmung der Moleküldimensionen. *Annalen der Physik* 19, 289–306. The equation is extremely simple: relative viscosity = $\eta/\eta_0 = 1 + 2.5\phi$, where ϕ represents volume fraction. The paper came from Einstein's miraculous period (1905–1906), while he was working in the federal patent office in Bern, during which time he also produced his theories of relativity and the photoelectric effect, and his paper on Brownian motion. The original publication in 1906 had an arith-

metical error in it (numerical constant equal to 1 instead of 2.5), which was corrected in 1911 in *Annalen der Physik* **34**, 591–592.

37. Arrhenius, S. (1917). The viscosity of solutions. *Biochemical Journal* **11**, 112–133. Arrhenius often does not live up to our image of a scholar who had been responsible for a major scientific revolution. Here he was almost petulant in his disdain for Einstein's equation.

38. Svedberg, T. and Sjögren, B. (1929). The molecular weight of Bence-Jones protein. *Journal of the American Chemical Society* **51**, 3594–3605.

39. Philpot, J. and Eriksson-Quensel, I.-B. were probably the first investigators to define the globular class clearly and to list some of the proteins that belong to it (pepsin, Bence-Jones protein, ovalbumin, etc.),[40] but they do not use the word 'globular'. W. T. Astbury originated the free use of the word and referred to Philpot and Eriksson-Quensel for definition of the criteria.[41]

40. Philpot, J. St L. and Eriksson-Quensel, I-B. (1933). An ultracentrifugal study of crystalline pepsin. *Nature* **132**, 932–933.

41. Astbury, W. T. and Lomax, R. (1934 and 1936). X-ray photographs of crystalline pepsin. *Nature* **133**, 795; **137**, 803.

42. Linderstrøm-Lang, K. (1924). Ref. 24.

43. Linderstrøm-Lang's own paper in 1924 contains some ambiguous language regarding the identification and location of charges, almost certainly the result of the author's youth and inexperience, which should not be allowed to detract from the credit due for the inspired decision to even think about using the Debye–Hückel theory as a model for analysis of protein titration curves. As was the practice of the time, the Ph.D. degree was awarded only after a substantial body of work had been published: in Linderstrøm-Lang's case it was in 1928 (see Holter, ref. 30, p. 97), four years after the 1924 paper.

44. Cohn, E. J. and Edsall, J. T. (1943). *Proteins, amino acids and peptides as ions and dipolar ions.* Reinhold, New York. ('Multipolar' might have been more accurate than 'dipolar'.)

45. Cohn, E. J. (1925). Ref. 25.

46. Cohn, E. J. and Prentiss, A. M. (1927). Ref. 1.

47. Cohn, E. J. (1932), Die Löslichkeitsverhältnisse von Aminosäuren und Eisweisskörpern: *Naturwissenschaften* **20**, 663–672.

48. Kirkwood, J. G. (1934). Theory of solutions of molecules containing widely separated charges with special application to zwitterions. *Journal of Chemical Physics* **2**, 351–361.

49. Tanford, C. and Kirkwood. J. G. (1957). Theory of protein titration curves. *Journal of the American Chemical Society* **79**, 5333–5339, 5340–5347.

50. Cohn, E. J. (1938). Number and distribution of the electrically charged groups of proteins. *Cold Spring Harbor Symposia on Quantitative Biology* **6**, 8–20.

Chapter 6

1. Polanyi, M. (1921). Summary of lecture: Die chemische Konstitution der Zellulose. *Naturwissenschaften* **9**, 288. Other materials given preliminary study in that year included serum albumin and haemoglobin in some kind of fibrous form. Silk fibroin came two years later.

2. Steinhardt, J., Fugitt, C. H., and Harris, M. (1940). Combination of wool protein with acid and base. *Journal of Research of the National Bureau of Standards* **24**, 335–367; **25**, 519–544.

3. For example, see the article on fibres in the 1911 edition of the *Encyclopaedia Britannica.*

4. Morris, P. J. T. (1986). *Polymer pioneers.* Center for History of Chemistry, Philadelphia.

5. Furakawa, Y. (1998). *Inventing polymer science.* Chemical Heritage Foundation, Philadelphia.

6. Morawetz, H. (1985). *Polymers.* John Wiley, New York.

7. Herzog (1878–1935) actually started his career as a biochemist, which may have been a factor in creating an easy community of interest There is a biography in *Neue Deutsche Biographie* **8**, 740–741, Munich 1953.

8. Bragg, W. H. and Bragg. W. L. (1913). The reflection of X-rays by crystals. *Proceedings of the Royal Society* **A 88**, 428–438.

9. Much of the early history is recounted, with biographical material and personal reminiscences of some of the persons involved, in Ewald, P. P. (ed.) (1962). *Fifty years of x-ray diffraction.* International Union of Crystallography, Utrecht. Chapter 5 of this book (pp. 57–80) provides a detailed history of the genesis of the Bragg equation.

10. Herzog, R. O. and Jancke, W. (1920). Über den physikalischen Aufbau einiger hochmolekularer organischen Verbindungen. *Berichte der deutschen chemischen Gesellschaft* **53**, 2162–2164.

11. Herzog, R. O., Jancke, W., and Polanyi, M. (1920). Röntgenspektrographische Beobachtungen an Zellulose. *Zeitschrift für Physik* **3**, 196–198, 343–348.

12. Polanyi was a fascinating, restless scientist. He went on from fibre crystallography (in Germany) to the theory of chemical kinetics (mostly in Manchester) and ultimately even became a professor of social sciences. For personal recollections of his three years in X-rays, see Ewald (ref. 9), pp. 629–636. For a general biography see Wigner, E. and Hodgkin, D. M. C. (1977). *Biographical Memoirs of Fellows of the Royal Society* **23**, 413–448.

13. Polanyi, M. (1921). Ref. 1.

14. Brill, R. (1923). Über Seidenfibroin. *Annalen der Chemie* **434**, 204–216. Brill suggested that silk fibroin was a mixture of a crystalline protein composed exclusively of alanine and glycine and an amorphous protein containing the other amino acids.

15. Sponsler, O. L. and Dore, W. H. (1926). The structure of Ramie cellulose as derived from x-ray data. *Colloid Symposium Monograph* **4**, 174–202. Although Sponsler and Dore were botanists, they obviously had a clear understanding of secondary bonds, presumably because they were at the University of California, where G. N. Lewis held sway and concepts like hydrogen bonds were first invented. They pointed out that only covalent bonds could provide the strength that was needed to create and maintain the long-range order that was observed within cellulose fibres. The German polymer scientists who had advocated colloidal association of small crystalline units in fibre formation had never addressed the problems of the nature of the bonds that would hold them together.

16. Meyer, K. H. and Mark, H. (1930). *Der Aufbau der hochpolymeren organischen Naturstoffe.* Akademische Verlagsgesellschaft, Leipzig.

17. For an excellent scientific biography, see Bernal, J. D., Astbury, W. T. (1963). *Biographical Memoirs of Fellows of the Royal Society* **9**, 1–35.

18. Meyer, K. H. and Mark, H. (1928). Über den Aufbau des Seiden-Fibroins. *Berichte der deutschen chemischen Gesellschaft* **61**, 1932–1936. Note that natural silk is virtually non-elastic; considerable confidence in the quantitation of the x-ray results was required to apply the silk structure to β-keratin.

19. Astbury, W. T. and Woods, H. J. (1930). The x-ray interpretation of the structure and elastic properties of hair keratin. *Nature* **126**, 913–914.

20. Astbury, W. T. and Street, A. (1930). X-ray studies of the structure of hair, wool, and related fibres (part I). *Philosophical Transactions of the Royal Society* **230**, 75–101.

21. Astbury, W. T. and Woods, H. J. (1933). X-ray studies of the structure of hair, wool and related fibres (part II). *Philosophical Transactions of the Royal Society* **232**, 333–394.

22. Watson. J. D. (1968). *The double helix.* Atheneum, New York.

23. Neurath, H. (1940). Intramolecular folding of polypeptide chains in relation to protein structure. *Journal of Physical Chemistry* **44**, 296–305.

24. Astbury, W. T. and Marwick, T. C. (1932). X-ray interpretation of the molecular structure of feather keratin. *Nature* **130**, 309–310.

25. For an overview see Astbury, W. T. (1936). X-ray studies of protein structure. *Nature* **137**, 803–805. Myosin was considered to be a single protein at the time, solely responsible for muscle elasticity.

26. Bernal, J. D. and Crowfoot, D. (1934). X-ray photographs of crystalline pepsin. *Nature* **133**, 794–795. Astbury himself had tried, but had been unable to get X-ray pictures for pepsin that displayed discrete spots. Bernal and Crowfoot ascribe this to insufficient care in handling the sample.

27. Astbury, W. T. (1934). X-ray studies of protein structure. *Cold Spring Harbor Symposia on Quantitative Biology* **2**, 15–23. See also: Astbury, W. T. (1937). Relation between 'fibrous' and 'globular' proteins. *Nature* **140**, 968–969.

Chapter 7

1. Tiselius, A. (1952). Presentation speech on the occasion of the award of a Nobel prize to A. J. P. Martin and R. L. M. Synge.

2. Proteid nomenclature, report of committee (1907). *Journal of Physiology* **35**, xvii–xx.

3. Recommendations of the committee on protein nomenclature (1908). *American Journal of Physiology* **21**, xxvii–xxx.

4. For the use of primary and secondary degradation products as direct evidence for high molecular weight as early as 1871, see p. 45 (ref.16 and note 17) in Chapter 4.

5. Reminiscence of A. J. P. Martin, in Ettre, L. S. and Zlatkis, A. (eds) (1979). *75 years of chromatography—a historical dialogue,* pp. 285–296. Elsevier, Amsterdam. To see this particular work in scientific context, see Martin A. J. P. and Synge, R. L. M. (1941). Separation of the higher monoamino-acids by counter-current liquid–liquid extraction: The amino-acid composition of wool. *Biochemical Journal* **35**, 91–121.

6. Biography by Pedersen, K. O. (1976). *Dictionary of Scientific Biography* **13**, 418–422. There is also an autobiographical account: Tiselius, A. (1968) Reflections from both sides of the counter. *Annual Review of Biochemistry* **37**, 1–24.

7. Pedersen, K.O. (1983). The Svedberg and Arne Tiselius. The early development of modern protein chemistry at Uppsala. In Semenza, G. (ed.), *Selected topics in the history of biochemistry: personal recollections* [vol. 35 of *Comprehensive Biochemistry*], pp. 235–256. Elsevier, Amsterdam.

8. Pedersen, K. O. (1983). Ref. 7, p. 265. The statement dates from 1934. A separation cell for the analytical ultracentrifuge was actually developed in 1937, with help from Tiselius.[9] Ultracentrifuges designed exclusively for mass separations were produced commercially around 1950.

9. Tiselius, A., Pedersen, K. O., and Svedberg, T. (1937), Analytical measurements of ultracentrifugal sedimentation. *Nature* **140**, 848–849 . This paper refers to preparative ultracentrifugation. The word 'analytical' in the title is intended to refer to biochemical analysis after separations, in contrast to optical analysis which takes place during separative flow. In the example cited, antibody activity was being measured. The tricky part of this business was to design a separation cell to stop contents of compartments from getting mixed up again when the centrifuge was slowed down from operating speed and stopped to permit samples to be taken.

10. Tiselius, A. (1937). A new apparatus for electrophoretic analysis of colloidal mixtures. *Transactions of the Faraday Society* **33**, 524–531. It is amusing to note (perhaps a last upsurge of what would soon become obsolete distinctions) that Tiselius first submitted his paper to a biochemical journal, but it was rejected for being 'too physical'.

11. Tiselius, A. (1937). Electrophoresis of serum globulin. *Biochemical Journal* **31**, 313–317, 1464–1477. The α, β, and γ notation was first used in the second segment of this report.

12. Picton and Linder (1892). Ref. 26, Chapter 4.

13. Compare comparable experimental difficulties in the development of the ultracentrifuge, as noted in Chapter 4, p. 58.

14. Zechmeister, L. (1946). Mikhail Tswett—The inventor of chromatography. *Isis* **36**, 108–109. For a later biography, see L. S. Ettre in Ettre, L. S. and Zlatkis, A. (eds) (1979). *75 years of chromatography—a historical dialogue*, pp. 483–490. Elsevier, Amsterdam.

15. A map of towns important in Tswett's life is provided in Ettre's biography (ref. 14). It includes Asti (Italy), where Tswett was born, in a hotel as it happened; Geneva, where he went to university; Warsaw, where he was associated successively with three different academic institutions and where he invented chromatography; Nizhnii Novgorod, site of a war-time academic job; Tartu (Estonia), where he held first full professorship; Voronezh, after the Germans took Tartu; Vladikavkaz (in the Caucasian mountains) for his health; and a dozen or so more.

16. Tswett, M. (1906) Physikalisch-chemische Studien über das Chlorophyll. Die Adsorption. *Berichte der deutschen botanischen Gesellschaft* **24**, 316–323.

17. Tswett, M. (1906) Adsorptionsanlyse und chromatographische Methode. Anwendung auf die Chemie des Chlorophylls. *Berichte der deutschen botanischen Gesellschaft* **24**, 384–393. References 16 and 17 have both been reprinted in English in Tswett, M. S. (trans. M. R. Masson) (1990). *Chromatographic adsorption analysis*. Ellis Horwood, New York.

18. Reminiscences of A. J. P. Martin and R. M. L. Synge, in Ettre, L. S. and Zlatkis, A. (eds) (1979). *75 years of chromatography—a historical dialogue*, pp. 285–296, 448–451. Elsevier, Amsterdam.

19. Martin, A. J. P. and Synge, R. L. M. (1945). Analytical chemistry of the proteins. *Advances in Protein Chemistry* **2**, 1–84. This review has more than 800 literature references.

20. Martin, A. J. P. and Synge, R. L. M. (1941). Analytical chemistry of the proteins. A new form of chromatogram employing two liquid phases. *Biochemical Jounal* **35**, 1358–1368.

21. Ref. 18, p. 295

22. Ion exchange chromatography was pioneered by Herbert Sober and Elbert Peterson at the National Institutes of Health. They used charged groups covalently linked to cellulose to create columns that would hold up either cations or anions. See Sober, H. A.

and Peterson, E. A. (1958). *Federation Proceedings* 17, 1116–1126. Sober and Peterson's article is part of a symposium on recent developments in separation methods that took place in Philadelphia in April 1958.[23]

23. Other papers at the symposium described the ultimate in mechanical complexity—the literal interpretation of 'partition' by use of a bank of 1000 serial separatory funnels (L. C. Craig and T. P. King, pp. 1126–1134)—and the ultimate in operational convenience, that would soon be generally adopted world-wide (S.Moore, D. H. Spackman, and W. H. Stein, pp. 1107–1115.)

24. Determann, H. (1969). *Gel chromatography*, 2nd edn. Springer, Berlin.

25. Porath, J. and Flodin, P. (1959). Gel filtration: a method for desalting and group separation. *Nature* 183, 1657–1659. The authors introduced cross-linked dextran gels, known today as Sephadex.

26. Lindquist, B. and Storgards, T. (1955). Molecular sieving properties of starch. *Nature* 175, 511–512. The first quantitative observations came accidentally, in the course of an examination of cheese extracts by zone electrophoresis. High-molecular weight peptides eluted fast; amino acids were retained in the starch support and eluted later.

27. The Bergmann–Niemann hypothesis will be cited among other visionary notions about protein structure in Chapter 10. It had a direct connection to amino acid analysis because it predicted a formula for the relative number of residues of individual amino acids in any protein molecule. The available evidence indicates that the hypothesis was not dismissed out of hand by analysts; Martin and Synge had deliberately included gelatin among the very first proteins they investigated[28] because the hypothesis was especially restrictive for this protein, which has an atypical overall composition.

28. Gordon, A. H., Martin, A. J. P., and Synge, R. L. M. (1941). A study of the partial acid hydrolysis of some proteins, with special reference to the mode of linkage of the basic amino acids. *Biochemical Journal* 35, 1369–1387.

29. Chemical, clinical and immunological studies on the products of human plasma fractionation. *Journal of Clinical Investigation* (1944). 23, 417–606. Note that the word used in the title is 'plasma', not 'serum'. Ten of the papers dealt with fibrinogen and other proteins involved in the formation of blood clots; 'serum' is by definition the clear fluid that remains after clots have been formed and removed.

30. Stein, W. H. and others (1946). Amino acid analysis of proteins. *Annals of the New York Academy of Sciences* 47, 57–239. There are no references to support Stein's later claim that Max Bergmann had been a source of inspiration.

31. Note the military persons on the fringes of the group picture in Fig. 5.3 (see p. 74). The famous textbook by Cohn, E. J. and Edsall, J. T. (1943). *Proteins, amino acids and peptides as ions and dipolar ions.* Reinhold, New York, which was for many years regarded as the most erudite book on protein science, was essentially completed before the intimate connection with the military was established.

32. Brand, E. (1946). Amino acid composition of simple proteins. *Annals of the New York Academy of Sciences* 47, 187–228.

33. The single letter symbols that are now standard for amino acid residues were introduced in 1966 to simplify computer searches for regularities within amino acid sequences. Dayhoff. M. O. and Eck, R. V. (1966). *Atlas of protein sequence and structure.* National Biomedical Research Foundation, Silver Spring, MD.

34. Martin, A. J. P. and Synge, R. L. M. (1941). Separation of the higher monoamino acids by counter-current liquid–liquid extraction: The amino acid composition of wool. *Biochemical Journal* 35, 91–121. Some years later, when they had become much better

known, the same authors provided a broad review of all applications to proteins, with more than 800 references. See ref. 21.

35. The method developed around 1900 to analyse for different types of nitrogen (Chapter 4, refs 39 and 45) had been adapted to analysis for lysine and arginine.

36. Snell, E. E. (1945). The microbiological assay of amino acids. *Advances in Protein Chemistry* 2, 85–118.

37. In addition to the methods we have explicitly mentioned, there existed a general isotope dilution method with potential applicability to all amino acids, but it was expensive to use and not widely adopted. Shemin, D. and Foster, G. L. (1946). The isotope dilution method of amino acid analysis. *Annals of the New York Academy of Sciences* 47, 119–134.

38. Edsall, J. T. (1946). Some correlations between physico-chemical data and the amino acid composition of simple proteins. *Annals of the New York Academy of Sciences* 47, 229–236.

39. H. T. Clarke was giving the concluding comments for the conference cited in ref. 30.

40. Moore, S. and Stein, W. H. (1949). Chromatography of amino acids on starch columns. Solvent mixtures for the fractionation of protein hydrolysates. *Journal of Biological Chemistry* 178, 53–77.

41. Moore, S. and Stein, W. H. (1949). Amino acid composition of β-lactoglobulin and bovine serum albumin. *Journal of Biological Chemistry* 178, 79–91.

42. Brand, E., Saidel, L. J., Goldwater, W. H., Kassel, B., and Ryan, F. J. (1945). The empirical formula of β-lactoglobulin. *Journal of the American Chemical Society* 67, 1524–1532.

Chapter 8

1. Hofmeister, F. (1908). Einiges über die Bedeutung und den Abbau der Eiweisskörper. *Archiv für experimentale Pathologie und Pharmakologie* (Supplement), 273–281. The quotation has been freely translated by C. T. and J. R.

2. Interview (1999). *Chemical Intelligencer* 5, no. 1, 8–14.

3. Brand, E., Saidel, L. J., Goldwater, W. H., Kassel, B., and Ryan, F. J. (1945). The empirical formula of β-lactoglobulin. *Journal of the American Chemical Society* 67, 1524–1532.

4. The DNP method and subsequent additions and variations are described in detail by Sanger F. (1952). The arrangement of amino acids in proteins. *Advances in Protein Chemistry* 7, 1–67.

5. Sanger, F. (1945). The free amino acids of insulin. *Biochemical Journal* 39, 507–515. This is where the DNFB method itself was first introduced.

6. Sanger, F. (1949). Fractionation of oxidized insulin. *Biochemical Journal* 44, 126–128.

7. Sanger, F. (1949). The terminal peptides of insulin. *Biochemical Journal* 45, 563–574. He concludes here that the intrinsic structural unit is 6000 not 12 000.

8. Sanger, F. and Tuppy, H. (1951). The amino acid sequence in the phenylalanine chain of insulin. *Biochemical Journal* 49, 463–481, 481–490.

9. It turned out that not enough overlapping bits of sequence were obtained from short-term acid hydrolysis and supplementary data from enzymic hydrolysis were required.

10. Sanger, F. and Thompson, E. O. P. (1953). The amino acid sequence in the glycyl chain of insulin. *Biochemical Journal* 53, 353–366, 366–374. As in the case of the phenylalanine chain, the first part of the report deals with products of partial acid hydrolysis and the second part with products of enzymic hydrolysis.

11. Ryle, A. P., Sanger, F., Smith, L. F., and Kitai, R. (1955). The disulphide bonds of insulin. *Biochemical Journal* **60**, 541–556.

12. Brown, H., Sanger, F., and Kitai, R. (1955). The structure of pig and sheep insulins. *Biochemical Journal* **60**, 556–565.

13. Pauling, L., Itano, H. A., Singer, S. J., and Wells, C. (1949). Sickle cell anemia. A molecular disease. *Science* **110**, 543–548.

14. Electrophoretic indication of genetic variants had been found in many common proteins. For example, two forms of β-lactoglobulin were observed. Some cows produced only one or the other, but some cows produced both. Also, many more haemoglobin variants were found, to add to the one derived from people with sickle-cell anaemia.

15. Colvin, J. R., Smith, D. B., and Cook, W. H. (1954). The microheterogeneity of proteins. *Chemical Reviews* **54**, 687–711.

16. Judson H. F. (1979). *The eighth day of creation. Makers of the revolution in biology*, pp. 213, 611. Jonathan Cape, London. Reprinted by Penguin Books, 1995, pp. 213, 611.

17. Crick, F. H. C. (1988). *What mad pursuits*, p. 34. Basic Books, New York.

Chapter 9

1. Edsall, J. T. (1986). Jeffries Wyman and myself: a story of two interacting lives. *Comprehensive Biochemistry* **36**, 99–195.

2. Adair, G. S. (1925). A critical study of the direct method of measuring the osmotic pressure of haemoglobin. *Proceedings of the Royal Society* **A108**, 627–637.

3. Adair, G. S. (1925). The osmotic pressure of haemoglobin in the absence of salts. *Proceedings of the Royal Society* **A109**, 292–300.

4. Hill, A. V. (1910). The possible effects of the aggregation of the molecules of haemoglobin on its dissociation curve. *Journal of Physiology* **40**, iv–vii.

5. Wyman, J. (1964). Linked functions and reciprocal effects in haemoglobin. A second look. *Advances in Protein Chemistry* **19**, 223–286.

6. A definitive historical account has been provided by Edsall, J. T. (1972). Blood and haemoglobin: the evolution of knowledge of functional adaptation in a biochemical system. *Journal of the History of Biology* **5**, 205–257.

7. Pedersen, K. O. (1983). See Chapter 7, ref. 7, p. 241.

8. Boyer, P. (1997). The ATP synthase—a splendid molecular machine. *Annual Review of Biochemistry* **66**, 717–749. The subunits are assembled into two distinct parts: a water-soluble part called F_1 and a membrane-bound part called F_0. The 55-kDal subunit is part of F_1, the 8-kDal subunit is part of F_0.

9. Blundell, T., Dobson, G., Hodgkin, D., and Mercola, D. (1972). Insulin: the structure in the crystal and its reflection in chemistry and biology. *Advances in Protein Chemistry* **26**, 274–402.

10. Steiner, D. F. and Oyer, P. E. (1967). The biosynthesis of insulin and a probable precursor of insulin by a human islet cell adenoma. *Proceedings of the National Academy of Sciences of the USA* **57**, 473–480; Steiner, D. F. and Clark, J. L. (1968). The spontaneous reoxidation of reduced beef and rat proinsulins. *Proceedings of the National Academy of Sciences of the USA* **60**, 622–629.

11. Melani. F., Rubenstein, A. H., Oyer, P. E., and Steiner, D. F. (1970). Identification of proinsulin and C-peptide in human serum by a specific immunoassay. *Proceedings of the National Academy of Sciences of the USA* **67**, 148–155. See also discussion in ref. 9, p. 372.

12. For a modern review of the fibrinogen < — > fibrin reaction, see Doolittle, R. F. (1975). *Advances in Protein Chemistry* **27**, 1–109. The most important enzyme in the fibrinogen/fibrin conversion is thrombin and it itself exists in the blood initially in the precursor form of prothrombin.
13. Northrop, J. H. (1935). The chemistry of pepsin and trypsin. *Biological Reviews* **10**, 263–282.
14. See Chapter 19.
15. Reid, K. B. M. and Porter, R. R. (1976). Subunit composition and structure of sub-component C1q of the first component of human complement. *Biochemical Journal* **155**, 19–23. The collagen-like segments are 78 residues long and the authors propose that they are the basis for a triple helix (such as exists in collagen itself), which binds three C1q subcomponents together into an overall trimeric structure.

Chapter 10

1. Astbury, W. T. (1934). X-ray studies of protein structure. *Cold Spring Harbor Symposia on Quantitative Biology* **2**, 15–23. This was only the second Cold Spring Harbor conference, less narrowly focused than they became later on. Astbury was the only protein person there.
2. See also Astbury, W. T. (1937). Relation between 'fibrous' and 'globular' proteins. *Nature* **140**, 968–969.
3. For an excellent biography, see Hodgkin, D. M. C. (1980). John Desmond Bernal. *Biographical Memoirs of Fellows of the Royal Society* **26**, 17–84. Among unusual personal details is Bernal's origins in Ireland, born on a farm near Tipperary.
4. Bernal, J. D. (1924). The structure of graphite. *Proceedings of the Royal Society* **A106**, 749–733.
5. Snow, C. P. (1934). *The search.* Macmillan, London.
6. Bernal, J. D. and Crowfoot, D. M. (1934). X-ray photographs of crystalline pepsin, *Nature* **133**, 794–795.
7. Northrop, J. H. (1930). Crystalline pepsin. I. Isolation and tests of purity. *Journal of General Physiology* **13**, 739–766.
8. An earlier more unusual wartime venture was the project HABBAKUK, aiming to construct huge unsinkable aircraft landing platforms from a mixture of woodpulp and ice.
9. A full biography of Dorothy Hodgkin has recently been published: Ferry, G. (1998). *Dorothy Hodgkin: a life.* Granta, London.
10. Bernal, J. D. and Crowfoot, D. M. (1933). Crystal structure of vitamin B_1 and of adenine hydrochloride. *Nature* **131**, 911–912.
11. Bernal, J. D. and Crowfoot, D. M. (1933). Crystalline phases of some substances studied as liquid crystals. *Transactions of the Faraday Society* **29**, 1032–1049.
12. Bernal, J. D., Crowfoot, D. M., and Fankuchen, I. (1940). X-ray crystallography and the chemistry of the sterols. *Philosophical Transactions of the Royal Society* **A239**, 135–182.
13. This was part of the work for which Hodgkin was awarded the Nobel Prize in Chemistry in 1964.
14. Svedberg, T. and Sjögren, B. (1929). The molecular weight of Bence-Jones protein. *Journal of the American Chemical Society* **51**, 3594–3605.
15. See Chapter 9.

16. A good place to catch the flavour of the Swedish enthusiasm for this idea is in the treatise by Svedberg and Pedersen[17], written in 1940 after enthusiasm elsewhere was pretty well dead. The number cited for the universal subunit molecular weight is usually 17 000, close to the molecular weight of a haeomoglobin subunit. See also Svedberg, T. (1929). Mass and size of protein molecules. *Nature* 123, 871; Svedberg, T. (1934). Sedimentation of molecules in centrifugal fields. *Chemical Reviews* 14, 1–15.

17. Svedberg, T. and Pedersen, K. O. (1940). *The ultracentrifuge.* Clarendon, Oxford. Reprinted in 1959 by the Johnson Reprint Corp., New York,

18. Bergmann M. and Niemann, C. (1937). On the structure of proteins: cattle hemoglobin, egg albumin, cattle fibrin, and gelatin. *Journal of Biological Chemistry* 118, 301–314.

19. The intrinsic fallacy of the claim of experimental support was exposed almost as soon as the hypothesis was first published. See Neuberger, A. (1939). Chemical criticism of the cyclol and frequency hypothesis of protein structure. *Proceedings of the Royal Society* A170, 64–65.

20. Fruton, J. S. (1999). *Proteins, enzymes, genes*, pp. 211–212. Yale University Press, Princeton, NJ.

21. There is a biographical article on Bergmann with a list of publications, written by B. Helferich, in *Chemische Berichte* (1969), 102, i–xxvi.

22. A full account of her tempestuous life is available: Abir-am, P. G. (1987). Synergy or clash: disciplinary and marital strategies in the career of mathematical biologist Dorothy Wrinch. In: Abir-am, P. G. and Outram, D. (eds). *Uneasy careers and intimate lives*, pp. 239–280. Rutgers University Press, New Brunswick, NJ. There are no indications that anti-feminine prejudice was ever involved in the criticisms of Wrinch's work which ultimately demolished her credibility.

23. Wrinch, D. M. (1936). The pattern of proteins. *Nature* 137, 411–412; Wrinch D. M. (1936). Energy of formation of 'cyclol' molecules. *Nature* 138, 241–242; Wrinch D. M. (1937). On the pattern of proteins. *Proceedings of the Royal Society* A160: 59–86; Wrinch D. M. (1937). The cyclol hypothesis and the 'globular' proteins'. *Proceedings of the Royal Society* A161, 505–524.

24. Wrinch, D. M. (1938). On the hydration and denaturation of proteins. *Philosophical Magazine* 25, 705–739.

25. Hager, T. (1995). *Force of nature: The life of Linus Pauling.* Simon & Schuster, New York.

26. Langmuir, I. (1937). Fundamental research and its human value. A paper given at 17th Congress of Applied Chemistry, Paris, Sept. 30, 1937. *GE Review* 40, 569–580.

27. Surface studies of proteins did in fact attract a small number of workers, but had only a minor impact and none at all on medical diagnosis.

28. Neurath, H. and Bull, H. B. (1938). The surface activity of proteins. *Chemical Reviews* 23, 391–435. See also Neuberger, loc. cit., ref. 19.

29. Langmuir, I. (1917). The constitution and fundamental properties of solids and liquids. II. Liquids. *Journal of the American Chemical Society* 39, 1848–1906.

30. Langmuir's biographer provides no clues, doesn't even mention Wrinch other than as a co-author in references. See Rosenfeld, A. (1966). *The quintessence of Irving Langmuir.* Pergamon Press, London.

31. Hodgkin, D. C, and Jeffreys, H. (1976). Obituary: Dorothy Wrinch. *Nature* 260, 564.

32. Mann, G. (1906). *Chemistry of the proteids.* Macmillan, London.

33. Anson, M. L. and Mirsky, A. E.(1925). On some general poperties of proteins. *Journal of General Physiology* 9, 169–179.

34. Anson, M. L. and Mirsky, A. E.(1931). The reversibility of protein coagulation. *Journal of Physical Chemistry* 35, 185–193.

35. These theoretical advantages are clearly spelled out by Anson, M. L. and Mirsky, A. E. (1929). Protein coagulation and its reversal. *Journal of General Physiology* 13, 121–132, 133–143; Anson, M. L. and Mirsky, A. E.(1934). Native and denatured hemoglobin. *Journal of General Physiology* 17, 393–408. The influential partnership between these two American investigators began when they were both doctoral students at Cambridge University in 1925. The collaboration was maintained when both went on to permanent positions at the Rockefeller Institute, despite the fact that they were at different branches of the institute, Mirsky at the main site (the hospital) in New York, Anson at the branch in Princeton. Anson was a co-founder in 1944 of protein science's scholarly review publication, *Advances in Protein Chemistry*.

36. Edsall, J. T. (1995). Hsien Wu and the first theory of protein denaturation (1931). *Advances in Protein Chemistry* 46, 1–5.

37. Wu, H. (1931). Studies on denaturation of proteins. XIII. A theory of denaturation. *Chinese Journal of Physiology* 5, 321–344. Reprinted in 1995 in *Advances in Protein Chemistry* 46, 6–26. It should be noted that Wu's 'theory' was based on several years of experimental studies, all virtually unknown in the western world. Wu, on the other hand, had access to and makes reference to applicable western work.

38. Pauling, L. (1993). Recollections: how my interest in proteins developed. *Protein Science* 2, 1060–1063.

39. Kauzmann, W. (1954). Denaturation of proteins and enzymes. In McElroy, W. D. and Glass, B. (eds), *The mechanism of enzyme action*, pp. 70–110. Johns Hopkins University Press, Baltimore.

40. Kauzmann, W. (1959). Some factors in the interpretation of protein denaturation. *Advances in Protein Chemistry* 14, 1–63.

41. Pace, C. N., Shirley, B. A., McNutt, M., and Gajiwala, K. (1996). Forces contributing to the conformational stability of proteins. *FASEB Journal* 10, 75–83.

42. The colloid aspects of textile materials and related topics (1933). 57th general discussion of the Faraday Society. *Transactions of the Faraday Society* 29, 1–368. See especially Speakman, J. B. and Hirst, M. C., The constitution of the keratin molecule, pp. 148–165, and approving comments from William Astbury in remarks made in the reported discussion.

43. Bernal, J. D. (1939). Structure of proteins. *Proceedings of the Royal Institution of Great Britain* 30, 541–557. Reprinted in *Nature* 143, 663–667.

44. Jacobsen, C. F. and Linderstrøm-Lang, K. (1949). Salt linkages in proteins. *Nature* 164, 411–412.

45. Anfinsen, C. B. (1972). Studies on the principles that govern the folding of protein chains. In *Nobel lectures in chemistry 1971–1980*, pp. 55–72. World Scientific Publishing, Singapore. This is a particularly well-written account of the relevant experiments and arguments. Anfinsen stresses the experiments he carried out under conditions where slow rearrangement of disulphide bonds was possible—they demonstrated that the correct disulphide bonds of active enzyme were spontaneously regenerated along with the refolding of the initially completely denatured and reduced polypeptide chain.

Chapter 11

1. Latimer, W. M. and Rodebush, W. H. (1920). Polarity and ionization from the standpoint of the Lewis theory of valence. *Journal of the American Chemical Society* 42, 1419–1433.

2. Lewis, G. N. (1916). The atom and the molecule. *Journal of the American Chemical Society* **38**, 762–785.

3. See also Langmuir, I. (1919). The arrangement of electrons in atoms and molecules. *Journal of the American Chemical Society* **41**, 868–934.

4. Bernal, J. D. and Fowler, R. H. (1933). A theory of water and ionic solution, with particular reference to hydrogen and hydroxyl ions. *Journal of Chemical Physics* **1**, 515–548.

5. Donohue, J. (1968). Selected topics in hydrogen bonding. In Rich, A. and Davidson, N. (eds), *Structural Chemistry and Molecular Biology*, pp. 443–465. W. H. Freeman, San Francisco.

6. Hodgkin, D. M. C. (1980). John Desmond Bernal. *Biographical Memoirs of Fellows of the Royal Society* **26**, 17–84.

7. Bernal, J. D. and Megaw, H. D. (1935). The function of hydrogen in intermolecular forces. *Proceedings of the Royal Society* **A151**, 384–420.

8. Laidler, K. J. (1993). *The world of physical chemistry*, pp. 200–202, 431–432. Oxford University Press.

9. Pauling, L. (1928). The shared-electron chemical bond. *Proceedings of the National Academy of Sciences of the USA* **14**, 359–362.

10. Pauling, L. (1939). *The nature of the chemical bond*, Chapter 9. Cornell University Press. 2nd ed., 1942. The material of interest to us is essentially unchanged in the two editions.

11. Pauling, L. (1993). Recollections. How my interest in proteins developed. *Protein Science* **2**, 1060–1063.

12. See Chapter 10.

13. Mirsky, A. E. and Pauling, L. (1936). On the structure of native, denatured, and coagulated proteins. *Proceedings of the National Academy of Sciences of the USA* **22**, 439–447.

14. Neurath, H., Greenstein, J. P., Putnam, F. W., and Erickson, J. O. (1944). The chemistry of protein denaturation. *Chemical Reviews* **34**, 157–265.

15. Astbury, see Chapter 6, p. 82.

16. Pauling, L., Corey, R. B., and Branson, R. H. (1951). The structure of proteins: two hydrogen-bonded helical configurations of the polypeptide chain. *Proceedings of the National Academy of Sciences of the USA* **37**, 205–210.

17. Pauling, L. and Corey, R. B. (1951). Configurations of polypeptide chains with favored orientations around single bonds: two new pleated sheets. *Proceedings of the National Academy of Sciences of the USA* **37**, 729–740.

18. For a definitive summary of all chemical work on synthetic polypeptides (mainly done at Courtaulds) see Bamford, C. H., Elliott, A., and Hanby, W. H. (1956). *Synthetic polypeptides*. Academic Press, New York.

19. Perutz, M. (1951). New x-ray evidence on the configuration of polypeptide chains. *Nature* **167**, 1053–1054.

20. Bragg, W. L., Kendrew, J. C., and Perutz, M. (1950). *Proceedings of the Royal Society* **A203**, 321–357.

21. It was also found in muscle; see Huxley, H. E. and Perutz, M. F. (1951). Polypeptide chains in frog sartorius muscle. *Nature* **167**, 1054. The 1.5-Å reflection is seen in both stretched and contracted muscle. Although the result confirms the α-helix, the principal emphasis is on the incompatibility with Pauling's speculation that an $\alpha <-> \beta$ transition is at the heart of the contraction mechanism.

22. Cochran, M. and Crick, F. H. C. (1952). Evidence for the Pauling–Corey α-helix in synthetic polypeptides. *Nature* **169**, 234–235.

23. Crick, F. H. C. (1952). Is α-keratin a coiled-coil? *Nature* 170, 882–883; Crick, F. H. C. (1953). The packing of α-helices: simple coiled-coils. *Acta Crystallographia* 6, 689–697.
24. Hager, T. (1995). *Force of nature: The life of Linus Pauling*, pp. 372–379. Simon & Schuster, New York. According to Hager, Pauling had an intense ambition to be acknowledged as the discoverer of the secret of protein structure. He and Corey were, of course, fully aware of the 5.4 Å/5.1 Å discrepancy, but Pauling decided to ignore it because 'he had to get into print first or risk losing his place in history'.
25. Pauling and Corey later proposed a seven-stranded coiled rope as basis for the 5.1-Å reflection,[26] but the suggestion did not have a significant impact. Structures for some other protein fibres (for example, collagen) proposed by Pauling at this time also proved to be based mainly on guesswork and incorrect.
26. Pauling, L., and Corey, R. B. (1953). Compound helical configurations of polypeptide chains, structure of proteins of the α-keratin type. *Nature* 171, 59–61.

Chapter 12

1. This chapter is based on an earlier article: Tanford, C. (1997). How protein chemists learned about the hydrophobic factor. *Protein Science* 6, 1358–1366.
2. Kauzmann, W. (1954). Denaturation of proteins and enzymes. In: McElroy, W. D. and Glass, B. (eds), *The mechanism of enzyme action*, pp. 70–110. Johns Hopkins University Press, Baltimore.
3. Kauzmann, W. (1959). Some factors in the interpretation of protein denaturation. *Advances in Protein Chemistry* 14, 1–63.
4. Kauzmann cites no earlier references, but disclaims originality and says that the idea had been 'in the air' when he wrote his reviews. (Kauzmann, personal communication, 1997.)
5. Edsall, J. T. (1985). Isidor Traube: physical chemist, biochemist, colloid chemist and controversialist. *Proceedings of the American Philosophical Society* 129, 371–406.
6. Traube, J. (1891). Ueber die Capillaritätsconstanten organischer Stoffe in wässriger Lösung. *Liebig's Annalen der Chemie* 265, 27–55.
7. Langmuir, I. (1917). The constitution and fundamental properties of solids and liquids. II. Liquids. *Journal of the American Chemical Society* 39, 1848–1906.
8. Hartley, G. S. (1936). *Aqueous solutions of paraffin-chain salts*. Hermann & Cie., Paris.
9. Gorter, E. and Grendel, E. (1925). On bimolecular layers of lipoids on the chromocytes of the blood. *Journal of Experimental Medicine* 41, 439–443.
10. This paper is now considered a great classic in the field of biology, but at the time it was ignored. The ludicrously slow acceptance of the bilayer concept for cell membranes (which took more than 40 years) has been described by one of the present authors in another place.[11]
11. Tanford, C. (1989). *Ben Franklin stilled the waves*. Duke University Press, Durham, NC.
12. Langmuir, I. (1938). Protein monolayers. *Cold Spring Harbor Symposia on Quantitative Biology* 6, 171–189.
13. Wrinch, D. M. (1937). On the structure of insulin. *Transactions of the Faraday Society* 33, 1369–13.
14. Crowfoot, D. (1938). The crystal structure of insulin. I. The investigation of air-dried crystals. *Proceedings of the Royal Society* A164, 580–602 .
15. Bernal, J. D. (1939). Vector maps and the cyclol hypothesis. *Nature* 143, 74–75.

16. Discussion on the protein molecules (1939). *Proceedings of the Royal Society* A170, 40–79.

17. Neuberger, A. (1939). Chemical criticism of the cyclol and frequency hypothesis of protein structure. *Proceedings of the Royal Society* A170, 64–65.

18. Hodgkin, D. M. C. (1980). John Desmond Bernal. *Biographical Memoirs of Fellows of the Royal Society* 26, 17–84.

19. Langmuir, I. (1938). The properties and structure of protein films. *Proceedings of the Royal Insitutution of Great Britain* 30, 483–496.

20. Langmuir, I. (1939). Pilgrim Trust Lecture. Molecular layers. *Proceedings of the Royal Society* A170, 1–39.

21. Langmuir, I. and Wrinch, D. (1939). Nature of the cyclol bond. *Nature* 143, 49–52.

22. Bernal, J. D. (1939). Ref. 15.

23. Bragg, W. L. (1939). Patterson diagrams in crystal analysis. *Nature* 143, 73–74.

24. Neville, E. H. (1938). Vector maps as positive evidence in crystal analysis. *Nature* 142, 994–995.

25. Abir-am, P. G. (1987). Synergy or clash: disciplinary and marital strategies in the career of mathematical biologist Dorothy Wrinch. In: Abir-am, P.G. and Outram, D. (eds), *Uneasy careers and intimate lives*, pp. 239–280. Rutgers University Press, New Brunswick, NJ.

26. Bernal, J. D. (1939). Structure of proteins. *Proceedings of the Royal Insitutution of Great Britain* 30, 541–557. Reprinted in *Nature* 143, 663–667.

27. Though Langmuir's own words are not used, proper credit is, of course, given to him as the original source of the idea.

28. Finding the gem of truth within the dross seems to have been a conscious habit with Bernal. Thus the following quotation is relevant, made with reference to some of Astbury's untenable ideas about protein structure: 'I found out early that when Astbury was talking it might appear to be nonsense but it always contained a valuable and new idea and I did my best at these meetings to interpret them' (See Bernal's biography of Astbury, Chapter 6, ref. 17)

29. Bernal, J. D. (1958). Introduction: Configurations and interactions of macromolecules and liquid crystals. *Discussions of the Faraday Society,* no. 25.

30. Frank, H. S. (1983). Citation classic: free volume and entropy in condensed systems. *Current Contents. Physical, Chemical & Earth Sciences* 23, no. 50, 22.

31. Frank, H. S. and Evans, M. W. (1945). Free volume and entropy in condensed systems. III. Entropy in binary liquid mixtures; partial molal entropy in dilute solutions; structure and thermodynamics in aqueous electrolytes. *Journal of Chemical Physics* 13, 507–532

32. Scheraga, H. A. (1960). Structural studies of ribonuclease. III. A model for the secondary and tertiary structure. *Journal of the American Chemical Society* 82, 3847–3852. Scheraga particularly favoured tyrosyl-carboxylate bonds as cross-linkers.

33. We return to the subject of the energetic driving force that leads to the native protein structure: what causes a linear polypeptide chain to fold into a compact globular entity? In the present chapter we have shown how the hydrophobic force was historically identified as being almost exclusively responsible, but in Chapter 10 (p. 119) we stated that at the end of the day hydrogen bonds as well as the hydrophobic force would be involved. The explanation is quite straightforward.

   1. It is not possible that hydrogen bonds between peptide groups (or other polar groups) can make a significant *favourable* contribution to the energetics of folding,

for essentially the reason given by J. D. Bernal in the 1939 lecture cited earlier: 'Ionic bonds are plainly out of the question,' he said, 'as they would certainly hydrate.' This statement applies to peptide groups as much as it does to ionic groups—they too will be hydrated when in solution as components of unfolded chains, with strong hydrogen bonds between protein and water. No significant energetic (thermodynamic) benefit can come from forming different hydrogen bonds of comparable strength within some tightly folded polypeptide structure.

2. However, it is easily conceivable that peptide groups might make an *unfavourable* contribution to the energetics of folding, simply by not participating in hydrogen bonding at all within the interior of the folded protein structure. The energetic advantage in hydrogen bonding to water molecules would then be lost and this would destabilize a folded structure, would neutralize the gain from the hydrophobic effect. Polypeptide chains in the interior of all native proteins in aqueous solution are in fact known to have a major fraction of their peptide groups in segments of α-helix or β-sheet. A purely arbitrary folded structure without internal hydrogen bonds would be unstable. (A corollary of this is that a few non-polar groups are invariably found sticking out of native protein molecules, making contact with water—the intrinsic energetic disadvantage can sometimes be overcome by a large gain from hydrogen bonds.)

The rules just outlined are somewhat modified for proteins destined to be bound to cell membranes, which themselves possess non-aqueous domains, but the underlying need for hydrogen bonds between 'buried' peptide groups remains the same. This is discussed in Chapter 19.

Chapter 13

1. Hodgkin, D. C. and Riley, D. P. (1968). Some ancient history of protein x-ray analysis. In Rich, A. and Davidson, N. (eds), *Structural chemistry and molecular biology* pp. 15–28.
2. Bernal, J. D. and Crowfoot, D. (1934). X-ray photographs of crystalline pepsin. *Nature* 133, 794–795.
3. Bernal, J. D. (1939). Structure of proteins. *Nature* 143, 663–667.
4. Patterson, A. L. (1935). A direct method for the determination of the components of interatomic distances in crystals. *Zeitschrift für Kristallographie* 90, 517–542. The article is in English.
5. Crowfoot, D. (1938). The crystal structure of insulin. I. The investigation of air-dried crystals. *Proceedings of the Royal Society* A164, 580–602.
6. Crowfoot, D. and Riley, D. (1938). An x-ray study of Palmer's lactoglobulin. *Nature* 141, 521–522. The molecular weight derived from cell dimensions was 36500.
7. Bernal, J. D., Fankuchen, I., and Perutz, M. (1938). An x-ray study of chymotrypsin and haemoglobin. *Nature* 141, 523–524.
8. Hägg, G. (1966). Presentation speech for the 1954 award of the Nobel prize to Linus Pauling. In *Nobel lectures—chemistry*. Elsevier, Amsterdam. Pauling's prize was awarded 'for his research into the nature of the chemical bond and its application to the elucidation of the structure of complex substances', not explicitly for his work related to proteins. The α-helix, still a brand-new development at the time, is mentioned, but with the caution that 'how far Pauling is right in detail still remains to be proved'.

9. Perutz, M. (1997). *Science is not a quiet life*. World Scientific, Singapore.

10. Perutz, M. F. (1947). A description of the iceberg aircraft carrier and the bearing of the mechanical properties of frozen wood-pulp upon some problems of glacier flow. *Journal of Glaciology* 1, 95–104. Another article on the project was written for a popular magazine and is reprinted in Perutz, M.F. (1989). *Is science necessary?* Barrie & Jenkins, London. For more information about Bernal's war-time efforts see Chapter 10.

11. Green, D. W., Ingram, V. M., and Perutz, M. F. (1954). The structure of haemoglobin IV. Sign determination by the isomorphous replacement method. *Proceedings of the Royal Society* **A225**, 287–307.

12. The MRC unit remained at the Cavendish Laboratory until after the intitially projected work of Perutz and Kendrew had in large part succeeded and even the relevant Nobel Prizes had been awarded. The present independent 'molecular biology laboratory' on the outskirts of the city was opened in 1962 to house at a single site not only the Perutz–Kendrew project, but also the research groups of Fred Sanger and of many others who had till then been scattered in different places around Cambridge.

13. Kendrew, J. C., Bodo, G., Dintzis, H. M., Parrish, R. G., Wyckoff, H., and Phillips, D. C. (1958). A three-dimensional model of the myoglobin molecule obtained by x-ray analysis. *Nature* **181**, 662–666.

14. Kendrew, J. C., Dickerson, R. E., Strandberg, B. E., Hart, R. G., Davies, D. R., Phillips, D. C., and Shore, V. C. (1960). Structure of myoglobin: A three-dimensional Fourier synthesis at 2 Å resolution. *Nature* **185**, 422–427.

15. Kendrew, J. C., Watson, H. C., Strandberg, B. E., Dickerson, R. E., Phillips, D. C., and Shore, V. C. (1961). The amino acid sequence of sperm whale myoglobin. A partial determination by x-ray methods, and its correlation with chemical data. *Nature* **190**, 666–672.

16. Kendrew, J. C. (1961). The three-dimensional structure of a protein molecule. *Scientific American* **205** (Dec.), 96–110.

17. Kendrew, J. C. (1963). Myoglobin and the structure of proteins. *Science* **139**, 1259–1266.

18. Perutz, M. F., Rossman, M. G., Cullis, A. F., Muirhead, H., Will, G., and North, A. C. T. (1960). Structure of haemoglobin: A three-dimensional Fourier synthesis at 5.5Å resolution, obtained by x-ray analysis. *Nature* **185**, 416–422.

19. Cullis, A. F., Muirhead, H., Perutz, M. F., Rossman. M. G., and North, A. C. T. (1962). Structure of haemoglobin: A three-dimensional Fourier synthesis at 5.5Å resolution. *Proceedings of the Royal Society* **A265**, 15–38, 161–187.

20. Bragg, L., Kendrew, J. C., and Perutz, M. F. (1950). Polypeptide chain configurations in crystalline proteins. *Proceedings of the Royal Society* **A203**, 321–357. The promise inherent in the greater simplicity of myoglobin was already pointed out at this time.

21. See Chapter 9, especially refs 4 and 5.

22. Perutz, M. F., Muirhead, H., Cox, J. M., and Goaman, L. C. G. (1968). Three-dimensional Fourier synthesis of horse oxyhaemoglobin at 2.8Å resolution: the atomic model. *Nature* **219**, 270–278.

23. Bolton, W. and Perutz, M. F. (1970). Three-dimensional Fourier synthesis of horse deoxy haemoglobiun at 2.8Å resolution. *Nature* **228**, 551–552.

24. Much earlier comparisons had been made between human reduced haemoglobin and horse oxyhaemoglobin, but only at 5.5 Å resolution. Muirhead, H. and Perutz, M. F. (1963). A three-dimensional Fourier synthesis of reduced human haemoglobin at 5.5 Å resolution. *Nature* **199**, 633–638. See also Perutz, M. F., Bolton, W., Diamond, R.,

Muirhead, H., and Watson, H. C. (1964). Structure of haemoglobin. An x-ray examination of reduced horse haemoglobim. *Nature* 203, 687–690.

25. Fermi, G., Perutz, M. F., Shaanan, S., and Fourme, R. (1984). The crystal structure of human deoxyhaemoglobin at 1.74 Å resolution. *Journal of Molecular Biology* 175, 159–174.

26. Perutz, M. F. (1964). The hemoglobin molecule. *Scientific American* 211 (Nov.), 64–76. Although an early review, this has particularly good graphic representations of the interaction betweem α and β subunits.

27. Perutz, M. F. (1976). Structure and mechanism of haemoglobin. *British Medical Bulletin* 32, 195–208.

28. Perutz, M. F. (1970). Stereochemistry of cooperative effects in haemoglobin. *Nature* 228, 726–734. This paper has good illustrations to demonstrate the nearly identical manner of folding for haemoglobins of different species.

29. Greenfield, N. and Fasman, G. D. (1969). Computed circular dichroism spectra for the evaluation of protein conformation. *Biochemistry* 8, 4108–4116.

30. Chapter 11, p. 130. Reference there is to Hager, T. (1995). *Force of nature: The life of Linus Pauling*. Simon & Schuster, New York.

31. Blake, C. C. F., Fenn, R. H., North, A. C. T., Phillips, D. C., and Poljak, R. J. (1962). Structure of lysozyme. A Fourier map of the electron density at 6 Å resolution obtained by x-ray diffraction. *Nature* 196, 1173–1176.

32. Phillips, D. C. (1966). The three-dimensional structure of an enzyme molecule. *Scientific American* 215 (Nov.), 78–90. The substrate for lysozyme is a polysaccharide from bacterial cell walls, which is destroyed in the lytic action of the enzyme.

33. Blake, C. C. F., Koenig, D. F., Mair, G. A., North, A. C. T., Phillips, D. C., and Sarma, V. R. (1965). Structure of hen egg-white lysozyme. A three-dimensional Fourier synthesis at 2 Å resolution. *Nature* 206, 757–761.

34. Johnson, L. N. and Phillips, D. C. (1965). Structure of some crystalline lysozyme-inhibitor complexes determined by x-ray analysis at 6 Å resolution. *Nature* 206, 761–763. Even at the low resolution (but with 2-Å map of protein alone available), it is possible to identify amino acid residues that may be unvolved in binding substrate. For higher resolution, see Blake, C. C. F., Johnson, L. N., Mair, G. A., North, A. C. T., Phillips, D. C., and Sarma, V. R. (1967). Crystallographic studies of the activity of hen egg-white lysozyme. *Proceedings of the Royal Society* B167, 378–388.

35. Blundell, T., Cutfield, J. F., Cutfield, S. M., Dobson, E. J., Dobson, G., Hodgkin, D., Mercola, D. M., and Vijayan, M. (1971). Atomic positions in rhombohedral 2-zinc insulin crystals. *Nature* 231, 506–511.

36. Results from several papers are reviewed (with excellent illustrations) by Blundell, T., Dobson, G., Hodgkin, D., and Mercola, D. (1972). Insulin: the structure in the crystal and its reflection in chemistry and biology. *Advances in Protein Chemistry* 26, 274–402.

Chapter 14

1. Loeb, J. (1898). The biological problems of today: physiology. *Science* 7, 154–156. The text is from a conference held in December 1897.

2. Mann, G. (1906). *Chemistry of the proteids, based on Professor Otto Cohnheim's 'Chemie der Eiweisskoerper'*. Macmillan, London. The text of the quotation has been slightly condensed.

3. *The Oxford dictionary of philosophy* (1994). Oxford University Press, Oxford.

4. Boyle, N. (1991). *Goethe: The poet and the age*. 3 vols. Oxford University Press, Oxford.

5. The modern *Dictionary of scientific biography* allots six pages to Goethe, describing his scientific fields as zoology, botany, geology, and optics. He was a competent amateur musician and also wrote an unpublished treatise on acoustics to supplement his work on optics.

6. A son of Fichte's was a friend of the family and a frequent visitor to the Helmholtz home when Helmholtz was a youth. Helmholtz obviously survived his early philosophical indoctrination. See Chapter 17 for a fuller account of his career.

7. Rechenberg, H. (1994). *Hermann von Helmholtz: Bilder seines Lebens und Wirkens.* VCH, Weinheim.

8. Lenoir, T. (1993). The eye as mathematician. In Cahan, D. (ed.), *Hermann von Helmholtz and the foundation of nineteenth-century science*, pp. 109–153. University of California Press.

9. Pauly, P. J. (1987). *Controlling life: Jacques Loeb and the engineering ideal in biology.* Oxford University Press.

10. In his early days, Hill contributed to the understanding of the physical chemistry of oxygen binding to haemoglobin. The 'Hill coefficient', an indicator of cooperativity in binding, is named after him. See Hill, A. V. (1910). The possible effects of the aggregation of the molecules of haemoglobin on its dissociation curve. *Journal of Physiology* **40**, iv–vii.

11. Bliss, M. (1982). *The discovery of insulin*. McClelland & Stewart, Toronto. Revised paperback edition (1988), Faber & Faber, London.

12. Holter, H. and Møller, M. (eds) (1976). *The Carlsberg Laboratories 1876/1976*. Rhodos, Copenhagen.

13. Kohler, R. E., Jr (1973). The enzyme theory and the origin of biochemistry. *Isis* **64**, 181–196.

14. Monod, J., Changeux, J.-P., and Jacob, F. (1963). Allosteric proteins and cellular control systems. *Journal of Molecular Biology*, **6**, 306–329.

15. Monod, J., Wyman, J., and Changeux, J-P. (1965). On the nature of allosteric transitions: a plausible mode. *Journal of Molecular Biology* **12**, 88–118.

16. Wyman, J. (1964). Linked functions and reciprocal effects in haemoglobin. A second look. *Advances in Protein Chemistry* **19**, 223–286.

17. Creagar, A. N. H. and Gaudillière, J. P. (1996). Meanings in search of experiments and vice versa. The invention of allosteric regulation in Paris and Berkeley, 1959–1968. *Studies in History and Philosophy of Biological and Biomedical Science* **27**, 1–89.

18. Massie, R. K. (1967). *Nicholas and Alexandra*. Dell, New York. This book describes the drama of the last years of the tsars, in which the haemophilia of the son and heir, Alexis, played a prominent part. The book contains a family tree, showing the inheritance of the disease via female carriers from Queen Victoria to the Russian, Spanish, and Prussian royal lines.

19. Chargaff, E. (1945). The coagulation of blood. *Advances in Enzymology* **5**, 31–65. This is a delightful early review, written before it was known that the overall process of blood clotting actually contains a host of proteolytic conversions between successive clotting 'factors'.

20. For modern reviews see Davie, E. W. and Fujikawa, K. (1975). Basic mechanisms in blood coagulation. *Annual Review of Biochemistry* **44**, 799–829. Also Jackson, C. M.

and Nemerson,Y. (1980). Blood coagulation. *Annual Review of Biochemistry* 49, 765–811. Both reviews catch the field at a moment when the outline of the cascade of factors is clear, but details are not yet complete.

21. Military interest in these proteins, used to stem the flow of blood on the battlefield, had a big influence on protein science during World War II. See Chapter 7.

22. Keilin, D. (1925). On cytochrome, a respiratory pigment, common to animals, yeast and higher plants. *Proceedings of the Royal Society* B98, 312–339. David Keilin (1887–1963) was a professor at Cambridge and right from the start recognized the existence of at least three spectrally different components in the system. In this paper, he also coined the term 'cytochrome', meaning 'cellular pigment'. The prosthetic group itself is modified in some of the cytochromes, but in three of them the same haem group is used as occurs in haemoglobin and myoglobin.

Chapter 15

1. Hoppe-Seyler, F. (1881). *Physiologische Chemie.* August Hirschwald, Berlin. Quoted by F. G. Hopkins.[2]

2. Hopkins, F. G., lecture given in 1913. See Needham, J. and Baldwin, E. (eds) (1949). *Hopkins and biochemistry,* p. 155. Heffer, Cambridge.

3. Hofmeister, F. (1901). *Die chemische Organization der Zelle.* Vieweg, Braunschweig.

4. de Réaumur, R. A. F. (1752). Observations sur la digestion des oiseaux. *Histoire de l'académie royale des sciences,* pp. 266, 461.

5. Berzelius, J. J. (1836). Einige Ideen über bei der Bildung organischer Verbindungen in der lebenden Natur wirksame, aber bischer nicht bemerkte Kraft. (Definition of catalysis.) *Jahres-Bericht über die Fortschritte der Chemie* 15, 237–245.

6. Fruton, J. S. (1972). *Molecules and life.* Wiley, New York. For an updated version, see Fruton, J. S. (1999). *Proteins, enzymes, genes.* Yale University Press, New Haven, CT.

7. Payen, A. and Persoz, J.-F. (1833). Mémoire sur la diastase, les principaux produits de ses reactions et leur applications aux arts industriels. *Annales de Chimie et de Physique* 53, 73–91.

8. Schwann, T. (1836) Über das Wesen des Verdaungsprocesses. *Archiv für Anatomie, Physiologie und wissenschaftliche Medizin* 90–138.

9. Wöhler, F. and Liebig, J. (1837), Über die Bildung der Bittermandelöls. *Annalen der Pharmacie* 22, 1–24.

10. Bernard, C. (1856). Mémoires sur le pancréas et sur le rôle du suc pancréatique dans le phénomènes digestifs, particulièrement dans la digestion de matières grasses neutres. *Supplément aux Comptes Rendus Hebdomadaires des Séances de l'Académie des Sciences* 1, 379–563. Bernard's discovery of lipase actually took place in 1846, ten years prior to the publication of his lengthy 'Mémoires'.

11. Berthelot, M. (1860). Sur la fermentation glucosique du sucre de canne. *Comptes Rendus Hebdomadaires des Séances de l'Académie des Sciences* 50, 980–984

12. Kühne, W. (1876). Ueber das Trypsin. *Verhandlungen des naturhistorisch-medizinischen Vereins zu Heidelberg* (Neue Folge) 1, 194–198

13. Bertrand, G. (1895). Sur la laccase et sur le pouvoir oxydant de cette diastase. *Comptes Rendus Hebdomadaires des Séances de l'Académie des Sciences* 120, 266–269

14. Kühne, W. (1876). Ueber das Verhalten verschiedener organisirter und sog. ungeformter Fermente. *Verhandlungen des Heidelberger naturhistorischen und medizinischen Verein* (Neue Folge) 1, 190–193. Both this paper and ref. 12 were reprinted in 1976 on

the occasion of the centenary of the first use of the word 'enzyme' in *FEBS Letters* **62**, suppl. 4.

15. Buchner, E. (1897). Alkoholische Gährung ohne Hefezellen. *Berichte der deutschen chemischen Gesellschaft* **30**, 117–124

16. Eduard Buchner was an organic chemist, trained, as had been Emil Fischer, by Alfred von Baeyer. For biographies, see Schriefers, H. (1970). *Dictionary of scientific biography* **2**, 560–563; Harries, C. (1917). *Berichte der deutschen chemischen Gesellschaft* **50**, 1843–1876.

17. Hill, A. C. (1898) Reversible zymohydrolysis. *Journal of the Chemical Society* **73**, 634–658. This was an exemplary paper, especially considering the date of 1898, when so much nonsense was still tolerated in relation to theoretical concepts like thermodynamics. Hill (1863–1947) went beyond merely demonstrating reversibility *per se*: he made quantitative measurements to show that the same equilibrium point was reached regardless of whether one started with maltose or glucose. The work was part of Hill's dissertation for a medical degree and he did nothing of a scholarly nature thereafter, but devoted himself to private medical practice.

18. Fischer, E. (1894). Einfluss der Konfiguration auf die Wirkung der Enzyme. *Berichte der deutschen chemischen Gesellschaft* **27**, 2985–2993; Fischer, E. (1909). *Untersuchen über Kohlenhydrate und Fermente*, Springer, Berlin (collected works from 1884 to 1908).

19. Oppenheimer, C. (1903). *Die Fermente und ihre Wirkungen*, 2nd edn Vogel, Lepzig. English translation of 1st edition in 1901: *Enzymes and their actions*. Charles Griffin, London.

20. Buckmaster, G. A. (1907). Behavior of blood and haematoporphyrin towards guaiaconic acid and aloin. Proceedings of the Physiological Society. *Journal of Physiology* **35**, 35–37.

21. Warburg, O. (1924). Ueber Eisen, den sauerstoffübertragenden Bestandteil des Atmungsferment. *Biochemische Zeitschrift* **152**, 479–494

22. Dony-Hénault, O. (1908). Contribution à l'étude méthodique des oxidases. 2e Mémoir. *Bulletin de l'Académie Royale de Belgique.* 105–163.

23. Cohnheim, O. (1912). *Enzymes.* Wiley, New York. The book contains the text of six lectures delivered at Johns Hopkins University in Baltimore.

24. Michaelis, L. (1909). Elektrische Überführung von Fermenten. I. Das Invertin. II: Trypsin and pepsin. *Biochemische Zeitschrift* **16**, 81–86, 486–488.

25. Michaelis, L. and Menten, M. L. (1913). Die Kinetik der Invertinwirkung. *Biochemische Zeitschrift* **49**, 333–369.

26. Oppenheimer, C. (1913). *Die Fermente und ihre Wirkungen*, 4th ed., Vogel, Lepzig. This edition, substantially expanded and revised, includes a chapter on physical chemistry by R. O. Herzog, at that time a member of the biochemistry faculty in Prague. Herzog later became director of the German institute for fibre research. (See Chapter 6.)

27. Willstätter, R. (1922). Ueber Isolieren von Enzymen. *Berichte der deutschen chemischen Gesellschaft* **55**, 3601–3623.

28. For a biography of Sumner, see Maynard, L. A. (1958). *Biographical Memoirs, National Academy of Sciences* **31**, 376–396.

29. Sumner, J. B. (1946). The chemical nature of enzymes. In *Les Prix Nobel en 1946*, pp. 185–192. Norstedt & Söner, Stockholm.

30. Sumner, J. B. (1926). The isolation and crystallization of the enzyme urease. Preliminary paper. *Journal of Biological Chemisty* **69**, 435–441.

31. Difficult to understand though it may be, the existence of hostility is indisputable. The

memory of it was still vivid in the minds of professional enzymologists a generation later. For example, the great fount of all wisdom in enzymology, a multi-volume serial publication edited by Paul Boyer and others (ref. 32), went into its second edition in 1959 with a dedication to the memory of J. B. Sumner, from which we quote the following:

> Willstätter and his students summarily dismissed Sumner's discovery with the 'explanation' that the protein was merely a carrier of the enzyme. When Willstätter rebuffed Sumner at a lecture on the Cornell campus, Sumner enlisted the aide of Gerty Cori to translate a manuscript so that it could be published in *Naturwissenschaften*. When Pringsheim, as the Baker non-residential lecturer, repeated Willstätter's performance, Sumner asked permission to present a lecture to Pringsheim's class and apparently convinced the latter of the validity of a crystalline enzyme protein.

These are events that took place at Sumner's own university (Cornell). His need to ask permission to defend his work here on his home ground is the best possible testimony to the intensity of feelings that it must have aroused.

32. Boyer, P. D., Lardy, H., and Myrbäck, K. (1959). *The enzymes*, 2nd ed., vol. 1. Academic Press, New York.
33. For a biography of Northrop see Herriott, R. M. (1962). A biographical sketch of John Howard Northrop. *Journal of General Physiology* 45, supplement, 1–16.
34. Northrop, J. H. (1930). Crystalline pepsin. Isolation and tests of purity. *Journal of General Physiology* 13, 739–767.
35. Northrop, J. H. (1935). The chemistry of pepsin and trypsin. *Biological Reviews* 10, 263–282.
36. Waldschmidt-Leitz, E. (1933). The chemical nature of enzymes. *Science* 78, 189.
37. Sumner, J. B. (1933). The chemical nature of enzymes. *Science* 78, 335.
38. Fruton, J. S. (1976). Biography of Richard Willstätter. *Dictionary of scientific biography* 14, 411–412.

Chapter 16

1. Ehrlich, P. (1897). Die Wertbemessung des Diphtherieheilserums und deren theoretische Grundlagen. *Klinisches Jahrbuch* 6, 299–326. The English translation, with the title 'The assay of the activity of diphtheria-curative serum and its theoretical basis', is from *The collected papers of Paul Ehrlich* (ref. 5).
2. Silverstein, A. M. (1989). The history of immunology. In W. E. Paul (ed.), *Fundamental immunology*, 2nd ed., Raven, New York.
3. Roux, E. and Yersin, A. (1888). Contribution à l'étude de la diphthérie. *Annales de l'Institut Pasteur* 2, 629–661.
4. von Behring, E. and Kitasato, S. (1890). Ueber das Zustandekommen der Diphtherie-immunität und der Tetanus-immunität bei Thieren. *Deutsche Medicinische Wochenschrift* no. 49, 1113–1114.
5. *The collected papers of Paul Ehrlich in four volumes* (ed. F. Himmelweit, M. Marquardt, and H. Dale) (1956–1960). Pergamon Press, London,
6. Metchnikoff, E. (1901). *Immunity in infectious diseases*. Macmillan, New York. Metchnikoff was an eccentric Russian, full of energy and originality. His personality is exceptionally well known by virtue of a biography written by his wife.[7]
7. Metchnikoff, O. (1921). *Life of Elie Metchnikoff*. Constable, London.

8. Rudolf Virchow (1821–1902) was the founder of pathology and obstinately believed that all pathology was due to cells going awry.
9. Ehrlich (1897), ref. 5 above.
10. Bordet, J. (1903). Sur le mode d'action des antitoxins sur les toxins. *Annales de l'Institut Pasteur* 17, 161–183.
11. Arrhenius, S. (1907). *Immunochemistry*. Macmillan, New York.
12. Landsteiner, K. (1936). *The specificity of serological reactions*. C. C. Thomas, Springfield, IL. The first edition (in German) was published in 1933. The most readily available edition in English is a much revised and extended reprint by Dover, New York, published in 1962.
13. Landsteiner, K. (1901). Ueber Agglutinationserscheinungen normalen menschlichen Blutes. *Wiener klinische Wochenschrift* 14, 1132–1134.
14. Tiselius, A. and Kabat, E. A. (1939). An electophoretic study of immune sera and purified antibody preparations. *Journal of Experimental Medicine* 69, 119–131.
15. Heidelberger, M., Pedersen, K., and Tiselius, A. (1936). Ultracentrifugal and electrophoretic studies on antibodies. *Nature* 138, 165. Heidelberger was a visiting fellow in Uppsala from the Presbyterian Hospital in New York. The γ-globulins he used were anti-pneumococcus antibodies. When taken from human or rabbit blood they had a molecular weight of 150 000, but the protein from horse had a molecular weight of 900000—that is it was an 'IgM' by the classification given in Table 16.1.
16. Neurath, H. (1939). Apparent shape of protein molecules. *Journal of the American Chemical Society* 61, 1841–1844.
17. Parventjev, I. A. (1936). US Patent 2065196.
18. M. L. Peterman studied the action of papain and was surprised to obtain quite different results from what she had expected on the basis of partial hydrolysis with trypsin or pepsin. Her fragments had a molecular weight of approximately 40 000. She herself did not investigate their binding activity, but later work[19,20] proved that they included the fully active 'Fab' fragment. Petermann, M. L. (1946). The splitting of human gamma globulin antibodies by papain and bromelin. *Journal of the American Chemical Society* 68,106–113.
19. Porter, R. R. (1958) Separation and isolation of fractions of rabbit gamma-globulin containing the antibody and antigenic combining sites. *Nature* 182, 670–671.
20. Porter, R. R. (1959). The hydrolysis of rabbit γ-globulin and antibodies with crystalline papain. *Biochemical Journal* 73, 119–126.
21. Edelman, G. (1959). Dissociation of γ-globulin. *Journal of the American Chemical Society* 81, 3155–3156.
22. Fleischman, J. B., Porter, R. R., and Press. E. M. (1963). The arrangement of the peptide chains of γ-globulin. *Biochemical Journal* 88, 220–228
23. Noelken, M. E, Nelson, C. A., Buckley, C. E., and Tanford, C. (1965). Gross conformation of rabbit 7S γ-immunoglobulin and its papain-cleaved fragments. *Journal of Biological Chemistry* 240, 218–224.
24. Valentine, R. C. and Green, N. M. (1967). Electron microscopy of an antibody–hapten complex. *Journal of Molecular Biology* 27, 615–617.
25. Huston, J. S., Margolies, M. N., and Haber, E. (1996). Antibody binding sites. *Advances in Protein Chemistry* 49, 329–450.
26. Poljak, R. J., Amzel, L. M., Avey, H. P., Chen, B. L., Phizackerly, R. P., and Saul, F. (1973). Three-dimensional structure of the Fab fragment of a human immunoglobulin at 2.8Å resolution. *Proceedings of the National Academy of Sciences of the USA* 70, 3305–3310.

27. Padlan, E. A. (1996). X-ray crystallography of antibodies. *Advances in Protein Chemistry* **49**, 57–133.

28. Kabat, E. A., Wu, T. T., Pery, H. M., Gottesman, K. S., and Foeller, C. (1991). *Sequences of proteins of immunological interest*, 5th ed., NIH Publication 91–3242. US Dept of Health and Human Services, Public Health Service, NIH, Bethesda, MD.

29. Breinl, F. and Haurowitz, F. (1930). Chemical investigation of the precipitate from hemoglobin and anti-hemoglobin serum and remarks on the nature of the antibodies. *Zeitschrift für physiologische Chemie* **192**, 45–57. (Translated title and reference citation were taken from *Chemical Abstracts.*)

30. Pauling, L. (1940). A theory of the structure and process of formation of antibodies. *Journal of the American Chemical Society* **62**, 2643–57.

31. Burnet, F. M. (1941). *The production of antibodies*, 1st ed., Macmillan, Melbourne.

32. Burnet, F. M. and Fenner, F. (1949). *The production of antibodies*, 2nd ed., Macmillan, Melbourne. The theoretical framework for information transfer was altered substantially.

33. Jerne, N. K. (1955). The natural selection theory of antibody formation. *Proceedings of the National Academy of Sciences of the USA* **41**, 849–857.

34. Niels Jerne's family had lived for many generations in Jutland, but he himself was born in London and lived and practised his profession in many different countries. He held joint appointments at the Danish State Serum Institute and the University of Copenhagen when his original work on the theory of antibody production was done. He was retired and living in Avignon at the time of his receiving a share of the Nobel Prize in 1984.

35. Burnet, F. M. (1957). A modification of Jerne's theory of antibody production using the concept of clonal selection. *Australian Journal of Science* **20**, 67–69.

36. Nossal, G. J. V. and Lederberg, J. (1958). Antibody production by single cells. *Nature* **181**, 1419–1420.

37. Talmage, D. W. (1959). Immunological specificity. *Science* **129**, 1643–1648.

38. This work was carried out independently in the laboratories of E. Haber and C. Tanford over several years. See Haber, E. (1964). Recovery of antigenic specificity after denaturation and complete reduction of disulfides in a papain fragment of anibody. *Proceedings of the National Academy of Sciences of the USA* **52**, 1099–1106; Whitney, P. L. and Tanford, C. (1965). Recovery of specific activity after complete unfolding and reduction of an antibody fragment. *Proceedings of the National Academy of Sciences of the USA* **53**, 524–532.

39. See concluding paragraph of Chapter 10.

40. Haber, E., ed. (1996). Antigen-binding molecules: antibodies and T-cell receptors. *Advances in Protein Chemistry* **49**, 1–536.

41. Bibel, D. J. (1988). *Milestones in immunology*. Science Tech, Madison, WI.

42. Sutton, B. J. and Gould, H. J. (1993). The human IgE network. *Nature* **366**, 421–8.

Chapter 17

1. Helmholtz, H. (1852). Ueber die Theorie der zusammengesetzten Farben. *Müller's Archiv für Anatomie, Physiologie, und wissenschaftlische Medizin* **46**, 1–482. English translation as Helmholtz, H. (1852). On the theory of compound colours. *Philosophical Magazine* (4th series) **4**, 519–534. (Text given here is slightly paraphrased from the original.)

2. From draft of a lecture on the theory of the three primary colours, given in 1861. Harman, P. M. (ed.) (1990). *The scientific letters and papers of James Clerk Maxwell*, vol. 1, pp. 675–679. Cambridge University Press.

3. Query 16 in Newton, I. (1789). *Opticks*, 4th ed, reprinted by Dover, New York, in 1952.

4. Helmholtz was only one of a number of pace-setters in physiology who began their careers under the leadership of Johannes Müller (1801–1858). The list included Theodor Schwann, one of the the founders of the cell theory, Emil DuBois-Reymond, the first major figure in electrophysiology, Rudolf Virchow, creator of scientific pathology, and others.

5. The quotation at the head of this chapter is a logical development of Müller's celebrated 'principle of specific nerve energies'. A clear and simple explanation of this principle in the context of the times is given in Charles Singer's classic general history of scientific ideas.[6]

6. Singer, C. (1959). *A short history of scientific ideas to 1900.* Oxford University Press.

7. Young, T. (1802). On the theory of light and colours, Bakerian lecture of the Royal Society. *Philosophical Transactions of the Royal Society* 92, 12–48. Young's three principal colours were red, yellow, and blue.

8. Young, one of the original proponents of the wave theory of light, was the first professor of natural philosophy at the newly created Royal Institution in London. Young considered himself primarily a physician and retained his professorship for only two years. His purely scientific publications were scanty because he did not want to jeopardize his prosperous London practice by giving the impression that his first loyalties lay elsewhere. Historical judgement has been that his comprehension of colour vision extended beyond what is explicitly stated in print. Maxwell and Helmholtz, who fifty years later published the definitive data on colour mixing (see below), both thought so and freely acknowledged Young's priority.

9. Maxwell, J. C. (1857). Experiments on colour, as perceived by the eye, with remarks on colour blindness. *Transactions of the Royal Society of Edinburgh* 21, 275–298; reprinted in Niven, W. D. (ed.) (1890). *The scientific papers of James Clerk Maxwell*, vol. 1, pp. 126–154. Cambridge University Press. An abstract, which includes a description of the colour top, was published in 1855 in *Proceedings of the Royal Society of Edinburgh* 3, 299–301; reprinted in Harman, P. M. (ed.) (1990). *The scientific letters and papers of James Clerk Maxwell*, vol. 1, pp. 287–289. Cambridge University Press.

10. Dalton, J. (1798). Extraordinary facts relating to the vision of colours; with observations. *Memoirs, Literary and Philosophical Society, Manchester,* no. 5. The common form of colour blindness is a sex-linked heritable defect, twenty times more frequent in men than women. This was certainly known to anyone in Dalton's time who took an interest in such matters and would for more than a century constitute a fixed point in theoretical genetics. The subject of colour vision as a whole must be seen as a potent argument for encouraging intellectual breadth in scientific endeavour.

11. Rushton, W. A. H. (1975). Visual pigments and color blindness, *Scientific American* 232 (March), 64–74.

12. Cahan, D. (ed.) (1993). *Hermann von Helmholtz and the foundation of nineteenth-century science.* University of California Press, Berkeley.

13. Rechenberg, H. (1994). *Hermann von Helmholtz: Bilder seines Lebens und Wirkens.* VCH, Weinheim.

14. Maxwell, J. C. (1876). Hermann von Helmholtz. *Nature* 15, 389–391.

15. Kurti, N. (1985). Helmholtz's choice. *Nature* 314, 499.

16. Helmholtz, H. loc. cit., ref. 1.

17. In the same vein, after the three-colour theory had been completely accepted, discrepancies between the precise λ_{max} values for each primary colour were reported by different laboratories and have been ascribed to differences in defining colour standards.

18. Sherman, P. D. (1981). *Colour vision in the nineteenth century*. Adam Hilger, Bristol. Helmholtz's overall work on vision is well presented by Lenoir, T. (1993). The eye as mathematician, in ref. 12, pp. 109–153, and by Kremer, R. L. (1993). Innovation through synthesis, ibid., pp. 205–258.

19. Rechenberg, H. (1994). loc. cit., ref. 13.

20. Schultze, M. (1866). Zur Anatomie und Physiologie der Retina. *Archiv für mikroskopische Anatomie* 2, 175–286.

21. See Geison, G. L. (1975). *Dictionary of scientific biography* 12, 230–233, for a biography and additional information.

22. Boll, F. (1877). Zur Anatomie und Physiologie der Retina. *Archiv für Anatomie und Physiologie (Physiologische Abteilung)* 4–35. The article is most readily accessible in an English translation, written on the occasion of the centenary in 1977, by Hubbard, R. *Vision Research* 17, 1253–1265. Boll (1849–1879) died at an early age; there are indications he might have pursued active recognition of his priority if he had lived longer.

23. Ewald, A. and Kühne, W. (1878). Untersuchungen über den Sehpurpur. *Untersuchungen des physiologischen Institut Heidelberg* 1, 44–66.

24. Kühne, W. (1879). Chemische Vorgänge in der Netzhaut. In *Handbuch der Physiologie* (L. Hermann, ed.) vol. 3, part 1. F. Vogel, Leipzig. English translation in 1977 by Hubbard, R. In *Vision Research* 17, 1273–1316. Also see Kühne, W. (1878). *On the photochemistry of the retina and on visual purple* (M. Foster, ed.). Macmillan, London.

25. Hecht, S. and Pickels, E. G. (1938). The sedimentation constant of visual purple. *Proceedings of the National Academy of Sciences of the USA* 24, 172–176. The measured molecular weight was much higher than what is now known to be the true value. Rhodopsin is a membrane protein and would have been in an aggregated state in this experiment.

26. Wald, G. (1935). Carotenoids and the visual cycle. *Journal of General Physiology* 19, 351–371.

27. Hecht, S. (1937). Rods, cones and the chemical basis of vision. *Physiological Reviews* 17, 239–289. Hecht (1892–1947) was the principal expert on the subject. In spite of the review's title, it avoids chemical specifics. Part of the review supposedly covers 'photochemistry', but the coverage is limited to principles, rate equations, etc., all for some hypothetical unnamed substance. For a somewhat later review by Hecht, see *The Harvey Lectures, 1937–38*, Williams & Wilkins, Baltimore. There is a good biography of Hecht by G. Wald, published in *Journal of General Physiology* 32, 1–16 (1948).

28. Wald, G. (1933). Vitamin A in the retina. *Nature* 132, 316–317. This is the first identification of a carotenoid with the visual process.

29. Wald, G. (1934). Carotenoids and the vitamin A cycle in vision. *Nature* 134, 65. One of the first positive identifications of rhodopsin as a protein. Wald concluded that the retinal must be protein-bound in its active state in dark-adapted retinas.

30. Wald, G. (1968). The molecular basis of visual excitation, *Nature* 219, 800–807. This is Wald's Nobel Prize lecture (1967) and an excellent historical summary. After his Nobel Prize, Ward donned casual clothes and joined the protest movement. He became the darling of the restless youth of the 1960s and tended to attract colourful audiences even to his formal academic lectures.

31. Marks, W. B., Dobelle, W. H., and MacNichol, E. F., Jr (1964). Visual pigments of single primate cones. *Science* 143, 1181–1183. λ_{max} values were at 445, 535 and 570 nm. Method of preparation of material was essentially the same as Schultze's.

32. Brown, P. K. and Wald, G. (1964).Visual pigments in single rods and cones of the human retina. *Science* 144, 45–52. They give the λ_{max} values as 450, 525, and 555 nm, compared to $\lambda_{max} = 505$ nm for rods.

33. Oxyhaemoglobin and deoxyhaemoglobin have significantly different spectra, but light absorption is not involved in molecular function. Ultraviolet spectra of the aromatic groups of tyrosine and tryptophan residues react to environmental change and can be used *as a tool* to follow the progress of protein denaturation, again with no direct relation to physiological function.

34. Nathans, J., Thomas, D., and Hogness, D. S. (1986). Molecular genetics of human color vision. *Science* 232, 193–202.

35. Nathans, J. (1989). The genes for color vision. *Scientific American* 260 (February), 28–35.

36. Kochendoerfer, G. G., Lin, S. W., Sakmar, T. P., and Mathies, R. A. (1999). How color visual pigments are tuned. *Trends in Biochemical Sciences* 24, 300–305.

Chapter 18

1. For a comprehensive biochemically oriented review of the history, see Needham, D. M. (1971). *Machina carnis*. Cambridge University Press.

2. Huxley, H. E. (1953). Electron microscope studies of the organisation of the filaments in striated muscle. *Biochimica et Biophysica Acta* 12, 387–394.

3. The Croonian lectures are still given regularly to this day, though the original instructions as to subject matter are interpreted rather freely.

4. Hoole, S. (translator) (1798). *Select works of A. van Leeuwenhoek*. Henry Fry, London.

5. Cuvier, G. (1805). *Leçons d'anatomie comparée*, vol. 1. Baudouin, Paris. The translated quotation is from ref. 1, pp. 28–29.

6. Helmholtz, H. (1847). Über die Erhaltung der Kraft, paper read before the German Physical Society on 23 July, 1847. Reprinted (1889) in Ostwald's *Klassiker der exacten Wissenschaften*. W. Engelmann, Leipzig.

7. Liebig, G. (1842). *Animal chemistry or organic chemistry in its application to physiology and pathology* (trans. W. Gregory), Cambridge University Press. Reprinted (1964) by the Johnson Reprint Corp., New York. We give a direct quotation on p. 17 (Chapter 1).

8. Kühne, W. (1859). Untersuchungen über Bewegungen und Veränderungen der contractilen Substanzen. *Archiv für Anatomie, Physiologie, und wissenschaftlische Medizin* 564–641, 748–835; Kühne, W. (1864). *Untersuchungen über das Protoplasma und die Contractilität*. W. Engelmann, Leipzig.

9. An outline of the actual preparative procedure was given in Chapter 2 to illustrate the crudity of preparative methods used at the time for non-crystalline materials. See p. 22.

10. Edsall, J. T. (1930). Studies in the physical chemistry of muscle globulin (myosin). *Journal of Biological Chemistry* 89, 289–313; von Muralt, A. and Edsall, J. T. (1930). Studies in the physical chemistry of muscle globulin. *Journal of Biological Chemistry* 89, 315–350, 351–386.

11. Engelhardt, V. A. and Lyubimova, M. N. (1939). Myosin and adenosinetriphosphatase. *Nature* 144, 668–669.

12. Lipmann, F. (1941). Metabolic generation and utilization of phosphate bond energy. *Advances in Enzymology* 1, 99–162.

13. Kalckar, H. M. (1941). The nature of energetic coupling in biological syntheses. *Chemical Reviews* 28, 71–178.

14. Kühne, W. (1888). On the origin and the causation of vital movement (Croonian Lecture). *Proceedings of the Royal Society* 44, 427–448.

15. Original reports of this and the following investigations were published as *Studies from the Institute of Medical Chemistry*, from the University of Szeged. Much of the work was republished under the name of A. Szent-Györgyi alone in *Acta Physiologica Scandinavica* 9, supplement XXV, 1–116 (1945), with full credit to the original reports included.

16. See also Szent-Györgyi, A. (1947). *Chemistry of muscular contraction.* Academic Press, New York. See ref. 18 for review of this book.

17. For a stirring account, see Szent-Györgyi, A. (1963). Lost in the twentieth century. *Annual Review of Biochemistry* 32, 1–14.

18. Bailey, K. (1947). Chemical basis of muscle contraction. *Nature* 160, 550–551. (Book review of the book by Szent-Györgyi which is referred to in ref. 16.) Bailey is someone who was almost exclusively a protein chemist. He had published an amino acid analysis of what was then called myosin in 1937; he spent a year in the Cohn–Edsall laboratory at Harvard in 1939. In 1946 (back in England) he discovered a third protein component of muscle, tropomyosin. There is a biography by A. C. Chibnall (1964) in *Biographical Memoirs of Fellows of the Royal Society* 10, 1–13.

19. F. B. Straub did not go to the US, but remained in Hungary. He became professor at the Semmelweiss medical school in Budapest, then became increasingly involved in public affairs. In 1988–1989, during Hungary's most recent years of political upheaval, he was actually the interim President of the State of Hungary!

20. Bailey, K. (1946). Tropomyosin: a new asymmetric protein component of muscle. *Nature* 157, 368–369.

21. Ebashi, S. (1963). Third component participating in the superprecipitation of 'natural actomyosin'. *Nature* 200, 1010; Ebashi, S., Ebashi, F., and Maruyama, K. (1964). A new protein factor promoting contraction of actomyosin. *Nature* 203, 645–646.

22. Maruyama, K., Matsubara, S., Natori, R., Nonomura, Y., Kimura, S., Ohashi, K., *et al.* (1977). Connectin, an elastic protein of muscle. *Journal of Biochemistry (Tokyo)* 82, 317–337 (1977); Wang, K. McClure, J., and Tu, A. (1979). Titin: major myofibrillar component of striated muscle. *Proceedings of the National Academy of Sciences of the USA* 76, 3698–3702.

23. Huxley, T. H. (1880). *The crayfish.* Kegan Paul, London.

24. Engelmann, T. W. (1895). On the nature of muscular contraction (Croonian Lecture at the Royal Society). *Proceedings of the Royal Society* 57, 411–433. Engelmann (1843–1909) was German, the son of the famous Leipzig publisher of books by Liebig, Kühne, and others, but his most productive years were spent as professor at the University of Utrecht. He was an exceptionally cultured man, a personal friend of Johannes Brahms, who dedicated a string quartet (no. 3 in B flat) to him in 1876.

25. Meyer, K. H. (1952). Thermoelastic properties of several biological systems. *Proceedings of the Royal Society* B139, 498–505. (The idea went back to a 1929 paper by Meyer!)

26. Weber, H. H. (1952). Is the contracting muscle in a new elastic equilibrium? *Proceedings of the Royal Society* B139, 512–520.

27. Hall, C. E., Jakus, M. A., and Schmitt, F. O. (1946). An investiugation of cross striations and myosin filaments in muscle. *Biological Bulletin* 90, 32–50.

28. Astbury, W. T. (1947). On the structure of biological fibres and the problem of muscle (Croonian Lecture). *Proceedings of the Royal Society* B134, 303–328.

29. Pauling, L. and Corey, R. B. (1951). The structure of hair, muscle and related proteins. *Proceedings of the National Academy of Sciences of the USA* 37, 261–271. This is an outrageously speculative proposal, in complete contrast to the preceding announcements of the α-helix and β-sheet structures themselves, which had followed a decade or more of rigorous groundwork. The proposal required wholesale rupture of hydrogen bonds, followed by reformation in an entirely different arrangement—and it had to be done reversibly. The authors themselves frankly admit: 'It is not so easy to suggest a single reasonable way in which the muscle can be reconverted to the stretched state.'

30. Pauling, L. and Corey, R. B. (1953). Compound helical configurations of polypeptide chains: structure of proteins of the α-keratin type. *Nature* 171, 59–61.

31. Astbury, W. T. (1950). X-ray studies of muscle. *Proceedings of the Royal Society* B134, 58–63.

32. Weber, H. H. and Portzehl, H. (1952). Muscle contraction and fibrous muscle proteins. *Advances in Protein Chemistry* 7, 161–252. The authors were from Tübingen, in what was then West Germany. Weber was one of the grandfathers of molecular muscle research, having been in the field since the 1930s.

33. Huxley, A. F. and Niedergerke, R. (1954). Structural changes in muscle during contraction. *Nature* 173, 971–973.

34. Huxley, H. and Hanson, J. (1954). Changes in the cross-striations of muscle during contraction and stretch and their structural interpretation. *Nature* 173, 973–976.

35. Huxley, H. E. (1996). A personal view of muscle and motility mechanisms. *Annual Review of Physiology* 58, 1–19.

36. Cohen, C. (1975). The protein switch of muscle contraction. *Scientific American* 233 (May), 36–45.

37. Wang, K. McClure, J., and Tu, A. (1979). Titin: major myofibrillar component of striated muscle. *Proceedings of the National Academy of Sciences of the USA* 76, 3698–3702; Trinick, J. (1996). Cytoskeleton: titin, a scaffold and spring. *Current Biology* 6, 258–260.

38. Keller, T. C. S. III (1997). Molecular bungees. *Nature* 387, 233–235.

39. Crick, F. H. C. (1952). The height of the vector rods in the three-dimensional Patterson of haemoglobin. *Acta Crystallographica* 5, 381–386. Pauling and Corey were aware of the inadequacy of the undistorted α-helix as a model for the α-keratin poteins. See ref. 30.

40. Crick, F. H. C. (1952). Is α-keratin a coiled-coil? *Nature* 170, 882–883.

41. Crick, F. H. C. (1953). The packing of α-helices: simple coiled-coils. *Acta Crystallographica* 6, 689–697.

42. Fletcher, W. M. and Hopkins, F. G. (1917), The respiratory process in muscle and the nature of muscular motion (Croonian Lecture). *Proceedings of the Royal Society* B89, 444–467.

43. Hill, A. V. (1932). The revolution in muscle physiology. *Physiological Reviews* 12, 56–67. Hill was still sceptical in 1949 about the identification of ATP as the energy supplier for contraction and he challenges the biochemists: 'Why not try to find out whether it really is, not in muscle extracts which cannot contract but in muscles which can?'[44] It does not seem to have occurred to him that he might himself do a testing experiment on intact muscle.

44. Hill, A. V. (1949). Adenosine triphosphate and muscular contraction. *Nature* 163, 320.

45. Hill, A. V. (1965). *Trails and trials in physiology*. Edward Arnold, London.

Chapter 19

1. Macallum, A. B. (1910). The inorganic composition of the blood in vertebrates and invertebrates, and its origin. *Proceedings of the Royal Society* B82, 602–624.
2. Fenn, W. O. (1940). The role of potassium in physiological processes. *Physiological Reviews* 20, 377–415. Fenn goes on to say that various explanations exist that can account for the scarcity of potassium in the sea, but that 'the reason for the great abundance of potassium in living cells is quite different and still unknown.'
3. Branden, C. and Tooze, J. (1999). *Introduction to protein structure*, 2nd edn. Garland, New York.
4. Hoppe-Seyler, F. (1881). *Physiologische Chemie*. August Hirschwald, Berlin.
5. Buchner, E. (1897). Alkoholische Gährung ohne Hefezellen. *Berichte der deutschen chemischen Gesellschaft* 30, 117–124.
6. For a minimal account, see Tanford, C. (1989). *Ben Franklin stilled the waves*. Duke University Press, Durham, NC. Overton was an expatriate, whose Ph.D. training and subsequent professional career were centred on the European continent. He published a remarkably detailed account of some of his experiments in Overton, E. (1902). Beiträge zur allgemeinen Muskel und Nervenphsiologie. *Pflüger's Archiv für die gesamte Physiologie* 92, 115–280, 346–386.
7. Singer, S. J. and Nicolson, G. L. (1972). The fluid mosaic membrane. *Science* 175, 720–731.
8. Spatz, L. and Strittmatter, P. (1971). A form of cytochrome b_5 that contains an additional hydrophobic sequence of 40 amino acids. *Proceedings of the National Academy of Sciences of the USA* 68, 1042–1046.
9. Tanford, C. and Reynolds, J. A. (1976). Characterization of membrane proteins in detergent solutions. *Biochimica et Biophysica Acta* 457, 133–170.
10. Helenius, A. and Simons, K. (1975). Solubilization of membranes by detergents. *Biochimica et Biophysica Acta* 415, 29–79.
11. Weber, K. and Osborn, M. (1969). The reliability of molecular weight determinations by dodecyl sulfate–polyacrylamide gel electrophoresis. *Journal of Biological Chemistry* 244, 4406–4412. The original proposal for the method came from Shapiro, A. L., Vinuela, E., and Maizel, J. V. (1967). *Biochemical and Biophysical Research Communications* 28, 815–820.
12. For physical definition of the state of polypeptide chains in SDS solution, essential theoretical underpinning for understanding the principles by which SDS gels work, see Reynolds, J. A. and Tanford, C. (1970). The gross conformation of protein–sodium dodecyl sulfate complexes. *Journal of Biological Chemistry* 245, 5161–5166.
13. Proteins in SDS were already being examined by electrophoresis in 1945 as part of an investigation to measure SDS binding. See Putnam, F. W. and Neurath, H. (1945). Interaction between proteins and synthetic detergents. *Journal of Biological Chemistry* 159, 195–209; 160, 397–408.
14. Trayer, H. R., Nozaki, Y., Reynolds, J. A., and Tanford, C. (1971). Polypeptide chains from human red blood cell membranes. *Journal of Biological Chemistry* 246, 4485–4488. This is one of several such analyses which quickly followed the availability of the SDS gel electrophoretic method. The original data were the familiar stained bands along the gel, but relative intensities were in this example quantitated by means of a spectrophotometric scan.
15. The discovery of the dominant role of ATP in bioenergetics is described in classic reviews published in 1941: Lipmann, F. (1941). Metabolic generation and utilization

of phosphate bond energy. *Advances in Enzymology* **1**, 99–162; Kalckar, H. M. (1941). The nature of energetic coupling in biological syntheses. *Chemical Reviews* **28**, 71–178.

16. Boyer, P. (1997). The ATP synthase—a splendid molecular machine. *Annual Review of Biochemistry* **66**, 717–749.

17. Branden, C. and Tooze, J. (1999). *Introduction to protein structure*, ref. 3. References to the original structural work are given.

18. Wilson, T. H. and Maloney, P. C. (1976). Speculations on the evolution of ion transport mechanisms. *Federation Proceedings* **35**, 2174–2179.

19. Hodgkin, A. L. and Huxley, A. F. (1939). Action potentials recorded from inside a nerve fibre. *Nature* **144**, 710–711; Hodgkin, A. L. and Huxley, A. F. (1945). Resting and action potentials in single nerve fibres. *Journal of Physiology* **104**, 176–185. These initial reports were followed by a series of papers in 1952 in *Journal of Physiology* **116**, 424–506; **117**, 500–504.

20. Skou, J. C. (1957). The influence of some cations on an adenosine triphosphatase from peripheral nerves. *Biochimica et Biophysica Acta* **23**, 394–401.

21. Skou, J. C. and Esmann, M. (1992). The Na,K-ATPase. *Journal of Bioenergetics and Biomembranes* **24**, 249–261.

22. Bernstein, J. (1902). Untersuchungen zur Thermodynamik der bioelektrischen Ströme. *Pflüger's Archiv für die gesamte Physiologie* **92**, 521–562.

23. Hodgkin, A. L. (1964). *The conduction of the nervous impulse*. Liverpool University Press.

24. Kuffler, S. W. and Nicholls, J. G. (1976). *From neuron to brain*. Sinauer, Sunderland, MA.

Chapter 20

1. Crick, F. H. C. (1958). On protein synthesis. *Symposia of the Society for Experimental Biology* **12**, 138–163.

2. Miescher, F. (1874). Das Protamin, eine neue organische Basis aus der Samenfaden des Rheinlacher. *Berichte der deutschen chemischen Gesellschaft* **7**, 376–379.

3. Hertwig, O. (1876). Beiträge zur Kenntnis der Bildung, Befruchtung und Theilung des Thierishen Eis. *Morphologisches Jahrbuch* **1**, 347–434.

4. Kossel, A. (1879). Ueber das Nuklein der Hefe. *Zeitschrift für physiologische Chemie* **3**, 284–291.

5. Weismann, A. (1893). *The germ-plasm: a theory of heredity*. Scribner's, New York.

6. See Chapter 8.

7. Judson, H. F. (1979). *The eighth day of creation. Makers of the revolution in biology.* Jonathan Cape, London. Reprinted by Penguin Books, 1995.

8. Watson, J. D. (1968). *The double helix*. Atheneum, New York.

9. Crick, F. H. C. (1988). *What mad pursuits*. Basic Books, New York.

10. Beadle, G. W. and Tatum, E. L. (1941). Genetic control of biochemical reactions in *Neurospora*. *Proceedings of the National Academy of Sciences of the USA* **27**, 499–506.

11. Beadle, G. W. (1945). Biochemical genetics. *Chemical Reviews* **37**, 15–96.

12. Beadle, G. W. (1946). Genes and the chemistry of the organism. *American Scientist* **34**, 31–53. (Bibliography concluded on p. 76.)

13. Muller, H. J. (1927). Artificial transmutation of the gene. *Science* **66**, 84–87.

14. Muller (1890–1967) moved to the University of Indiana in 1945; he was one of J. D. Watson's teachers there.

15. Delbrück, M. (1940). Radiation and hereditary mechanims. *American Naturalist* **74**, 350–362.

16. Stanley, W. M. and Knight, C. A. (1941). The chemical composition of strains of tobacco mosaic virus. *Cold Spring Harbor Symposia on Quantitative Biology* **9**, 255–262. TMV was rod-shaped, with an overall molecular weight of about 50 million.

17. Muller, H. J. (1941). Résumé and perspectives of the symposium on genes and chromosomes. *Cold Spring Harbor Symposia on Quantitative Biology* **9**, 290–308.

18. Haldane, J. B. S. (1942). *New paths in genetics*. Harper, New York.

19. Svedberg, T. (1939). Opening addresss, A discussion on the protein molecule. *Proceedings of the Royal Society* **A170**, 40–56 (1939). The occasion was devoted to protein *structure*. There was no obligation for Svedberg to speculate about genes; he must have been eager to display his awareness of the consensus.

20. Pauling, L. and Corey, R. B. (1953). A proposed structure for the nucleic acids. *Proceedings of the National Academy of Sciences of the USA* **39**, 84–97.

21. Hager, T. (1995). *Force of nature: the life of Linus Pauling*, p. 396. Simon & Schuster, New York.

22. To quote directly from the introduction to the paper, 'An understanding of the molecular structure of the nucleic acids should be of value in the effort to understand the fundamental phenomena of life'. The paper did not specify how.

23. Avery, O. T., MacLeod, C. M., and McCarty, M. (1944). Studies on the nature of the substance inducing transformation of pneumococcal types. Induction of transformation by a desoxyribonucleic acid fraction isolated from Pneumococcus Type III. *Journal of Experimental Medicine* **79**, 137–158.

24. For example, Judson, H. F., ref. 7, and, more recently, Fruton, J. S. (1999). *Proteins, enzymes, genes*. Yale University Press, New Haven, NC. The latter says: 'As I read the historical record, that inability [acceptance of Avery's result], at least in the case of Delbrück and Luria, came from their low opinion of DNA as the [possible] bearer of genetic factors.'

25. Hunter, G. K. (1999). Phoebus Levene and the tetranucleotide structure of nucleic acids. *Ambix* **46**, 73–103. (Note that Levene and Avery were colleagues at the Rockefeller Insitute.)

26. Delbrück, M. (1941). A theory of autocatalytic synthesis of polypeptides and it application to the problem of chromosome reproduction. *Cold Spring Harbor Symposia on Quantitative Biology* **9**, 122–124.

27. Pauling, L. and Delbrück, M. (1940). The nature of the intermolecular forces operative in biological processes. *Science* **92**, 77–79.

28. Delbrück, ref. 15. For a good account of Delbrück's career (with a full list of publications), see Hayes, W. (1982). Biography of Max Delbrück. *Biographical Memoirs of Fellows of the Royal Society* **28**, 58–90, and the references cited therein.

29. Francis Crick has commented in his personal history on the pre-eminence of chemical structures in molecular biology. He describes the true physicist, Max Delbrück, as follows: 'I don't think Delbrück much cared for chemistry. Like most physicists, he regarded chemistry as a rather trivial application of quantum mechanics.' See Crick's *What mad pursuits* (ref. 9), p. 61.

30. Moore, W. (1989). *Schrödinger: life and thought*. Cambridge University Press.

31. Schrödinger, E. (1946). *What is life?*, p. 76. Cambridge University Press. The book is based on lectures delivered at Trinity College, Dublin, in February 1943.

32. Perutz, M. F. (1987). Physics and the riddle of life. *Nature* **326**, 555–559.

33. See biography of Gamow by Stuewer, R. H. (1972). *Dictionary of scientific biography* 5, 271–273.

34. Hershey, A. D. and Chase, M. (1952). Independent functions of viral protein and nucleic acid in growth of bacteriophage. *Journal of General Physiology* 36, 39–56.

35. Judson, ref. 7, pp. 130–131.

36. Chargaff, E. (1950). Chemical specificity of nucleic acids and mechanism of their enzymic degradation. *Experientia* 6, 201–209.

37. Watson, J. D. and Crick, F. H. C. (1953). Molecular structure of nucleic acids. A structure for deoxyribose nucleic acid. *Nature* 171, 737–738.

Chapter 21

1. Crick, F. H. C. (1966). The genetic code III. *Scientific American* 215 (Oct.), 55–62.

2. Nirenberg, M. W., Jones, O. W., Leder, P., Clark, B. F. C., Sly, W. S., and Pestka, S. (1963). On the coding of genetic information. *Cold Spring Harbor Symposia on Quantitative Biology* 28, 549–557. A scientific review from the people who did more than anyone else to make it possible.

3. Nirenberg, M. W. (1963). The genetic code. II. *Scientific American* 208 (Mar.), 80–94. A review directed at the non-expert, with lots of pictures to represent molecular events, as well to show equipment used in the lab.

4. Crick, F. H. C. (1958). On protein synthesis. *Symposia of the Society for Experimental Biology* 12, 138–163. This was part of the 12th symposium of the Society for Experimental Biology, entitled 'The Biological Replication of Macromolecules'.

5. See preceding chapter, p. 233 and ref. 33.

6. Gamow, G. (1954). Possible relation between deoxyribonucleic acid and protein structure. *Nature* 173, 318.

7. Crick, F. H. C., Barnett, L., Brenner, S., and Watts-Tobin, R. J. (1961). General nature of the genetic code for proteins. *Nature* 192, 1227–1232.

8. For a history with emphasis on the biochemical details and not on the code itself, see Zamecnik, P. (1969). An historical account of protein synthesis, with current overtones—A personalized view. *Cold Spring Harbor Symposia on Quantitative Biology* 34, 1–14.

9. Many academic departments of biochemistry have become departments of 'biochemistry and molecular biology'. The major professional association for biochemists in America, the American Society for Biochemistry, changed its name to the American Society for Biochemistry and Molecular Biology in 1987.

10. Cellular Regulatory Mechanisms (1961). *Cold Spring Harbor Symposia on Quantitative Biology* 26, 1–408.

11. Synthesis and Structure of Macromolecules (1963). *Cold Spring Harbor Symposia on Quantitative Biology* 28, 1–610.

12. The genetic code (1966). *Cold Spring Harbor Symposia on Quantitative Biology* 31, 1–762.

13. Grunberg-Manago, M. and Ochoa, S. (1955). Enzymatic synthesis and breakdown of polynucleotides. Polynucleotide phosphorylase. *Journal of the American Chemical Society* 77, 3165–3166. This work demonstrates that reversible phosphorylation may be a major mechanistic step in breakdown or synthesis.

14. Hoagland, M. B., Keller, E. B., and Zamecnik, P. C. (1956). Enzymatic carboxyl activation of amino acids. *Journal of Biological Chemistry* 218, 345–358.

15. Hoagland, M. B., Zamecnik, P. C., and Stephenson, M. L. (1957) Intermediate reactions in protein biosynthesis. *Biochimica et Biophysica Acta* **24**, 215–216.
16. Zamecnik, P. C., Keller, E. B., Littlefield, J. W., Hoagland, M. B., and Loftfield, R. B. (1956). Mechanism of incorporation of labeled amino acids into proteins. *Journal of Cellular and Comparative Physiology* **47**, suppl. 1, 81–101.
17. Jacob, F. and Monod, J. (1961). Genetic regulatory mechanisms in the synthesis of proteins. *Journal of Molecular Biology* **3**, 318–356.
18. Brenner, S., Jacob, F., and Meselson, M. (1961). An unstable intermediate carrying information from genes in the nucleus to ribosomes for protein synthesis. *Nature* **190**, 576–581.
19. Brenner, S. (1961). RNA, ribosomes and protein synthesis. *Cold Spring Harbor Symposia on Quantitative Biology* **26**, 101–110.
20. Nirenberg, M. and Matthaei, J. (1961). The dependence of cell-free protein synthesis in *E. coli* upon naturally occurring or synthetic polyribonucleotides. *Proceedings of the National Academy of Sciences of the USA* **47**, 1588–1602.
22. Matthaei, J. H., Jones, O. W., Martin, R. G., and Nirenberg, M. (1962). Characteristics and composition of RNA coding units. *Proceedings of the National Academy of Sciences of the USA* **48**, 666–677.
22. Speyer, J. F. Lengyel, P., Basilio, C., and Ochoa, S. (1962). Synthetic polynucleotides and the amino acid code. II. *Proceedings of the National Academy of Sciences of the USA* **48**, 63–68. Several more papers followed in quick succession to help complete the assignment of code words.
23. Khorana. H. G. and others (1966). Polynucleotide synthesis and the genetic code. *Cold Spring Harbor Symposia on Quantitative Biology* **31**, 39–49. For description of the final phases of the underlying original work see Khorana and co-workers (1965). Studies on polynucleotides XLIII to XLVI. *Journal of the American Chemical Society* **87**, 2956–2970, 2971–2981, 2981–2988, 2988–2995.
24. Leder, P. and Nirenberg, M. W. (1964). RNA codewords and protein synthesis, II. Nucleotide sequence of a valine RNA codeword. *Proceedings of the National Academy of Sciences of the USA* **52**, 420–427.
25. Nirenberg, M. W. and Leder, P. (1964). RNA codewords and protein synthesis. *Science* **145**, 1399–1407. The paper had a more explicit subtitle: 'The effect of trinucleotides upon the binding of sRNA to ribosomes'.

Chapter 22

1. Johnson, O. B. (1928). *A study of Chinese alchemy*. Commercial Press, Shanghai.
2. Crick, F. H. C. (1958). On protein synthesis. *Symposia of the Society of Experimental Biology* **12**, 138–163; the quotation comes from p. 142.
3. Chothia, C. (1992). One thousand families for the molecular biologist. *Nature* **357**, 543–544.
4. Margoliash, E. and Schejter, A. (1967). Cytochrome *c*. *Advances in Protein Chemistry* **21**, 113–286. The evolution of the protein's amino acid sequence is covered on pp. 175–205.
5. Dickerson, R. E. (1971). The structure of cytochrome *c* and the rates of molecular evolution. *Journal of Molecular Evolution* **1**, 26–45.
6. Dickerson, R. E. (1972). The structure and history of an ancient protein. *Scientific American* **226** (4), 58–72.

7. Pauling, L., Itano, H. A., Singer, S. J., and Wells, C. (1949). Sickle cell anemia. A molecular disease. *Science* **110**, 543–548.

8. Persons with 'sickle-cell trait', a more common disorder than sickle-cell anaemia, in which no serious pathological consequences are ordinarily observed, were found to have a mixture of normal and abnormal proteins in their blood: electrophoretic scans of their haemoglobin were the same as those obtained for a mixture of pure normal and variant proteins. This was one of several control experiments that were performed to establish the conclusions beyond any reasonable doubt.

9. Ingram, V. M. (1956). A specific chemical difference between the globins of normal and sickle-cell anaemia haemoglobins. *Nature* **178**, 792–794.

10. Ingram, V. M. (1957). Gene mutations in human haemoglobin: the chemical difference between normal and sickle cell haemoglobin. *Nature* **180**, 326–328.

11. We are quoting Ingram's exact words. In sickle-cell trait, where the abnormal protein is diluted 1:1 by normal haemoglobin, it is equally understandable that ease of crystallization would be significantly less enhanced and would not lead to a significant medical problem except under conditions of extreme oxygen deprivation.

12. Itano, H. A. (1957). The human hemoglobins: their properties and genetic control. *Advances in Protein Chemistry* **12**, 216–269

13. Branden, C. and Tooze, J. (1999). *Introduction to protein structure*, 2nd edn. Garland, New York. Chapter 17 is entitled 'Prediction, engineering, and design of protein structures'.

14. DeGrado, W. F. (1997). Proteins from scratch. *Science* **278**, 80–81.

15. Dahiyat, B. I. and Mayo, S. L. (1997). De novo protein design: fully automated sequence selection. *Science* **278**, 82–87.

Subject index

Name index